Mathematical Reasoning with Diagrams

Mathematical Reasoning with Diagrams

From Intuition to Automation

Mateja Jamnik

Center for the Study of
Language and Information
Stanford, California

Center for the Study of Language and Information
Leland Stanford Junior University
Printed in the United States
05 04 03 02 01   5 4 3 2 1

Library of Congress Cataloging-in-Publication Data

Jamnik, Mateja, 1973–
Mathematical reasoning with diagrams : from intuition to automation /
Mateja Jamnik.
p. cm. — (CSLI lecture notes ; no. 127)
Includes bibliographical references and index.

ISBN 1-57586-323-5 (cloth : alk. paper)
ISBN 1-57586-324-3 (pbk : alk. paper)

1. Mathematics–Charts, diagrams, etc.
2. Logic, Symbolic and mathematical.
3. Automatic theorem proving.
I. Title. II. Series.
QA90.J33 2001
511.3—dc21
2001047923
CIP

∞ The acid-free paper used in this book meets the minimum requirements of the American National Standard for Information Sciences—Permanence of Paper for Printed Library Materials, ANSI Z39.48-1984.

CSLI was founded early in 1983 by researchers from Stanford University, SRI International, and Xerox PARC to further research and development of integrated theories of language, information, and computation. CSLI headquarters and CSLI Publications are located on the campus of Stanford University.

CSLI Publications reports new developments in the study of language, information, and computation. In addition to lecture notes, our publications include monographs, working papers, revised dissertations, and conference proceedings. Our aim is to make new results, ideas, and approaches available as quickly as possible. Please visit our web site at http://cslipublications.stanford.edu/ for comments on this and other titles, as well as for changes and corrections by the author and publisher.

Contents

Foreword

The advent of the modern computer in the nineteen-fifties immediately suggested a new research challenge: to seek ways of programming these versatile machines which would make them behave as much like intelligent human beings as possible. After fifty years or so, this quest has produced some intriguing results, but until now progress has been disappointingly slow. This book is a welcome and encouraging sign that things may at last be about to change.

To be sure, what might be called the *purely logical* approach has recently produced some noteworthy successes. Consider the following two examples:

On October 10, 1996, a rigorous proof of the Robbins Conjecture was found by William McCune's theorem proving program EQP at the Argonne National Laboratory. This problem had been unsolved since the mid-nineteen-thirties.

On May 11, 1997, the (then) world chess champion Garry Kasparov lost a six game match against the computer program Deep Blue with a score of 2.5 to 3.5: two wins for Deep Blue, one win for Kasparov and three draws.

Both these programs, however, use quite *non-human* methods. Neither of them is at all based on how the mind of the human expert actually works. It is in fact very difficult to find out what the natural intellectual processes of expert humans really are. To program a computer to solve the kind of problems that such experts can solve, the *purely logical* approach has hitherto been found more effective than the *heuristic* approach: to invent systematic algorithms for solving the problems rather than trying to discover, and then to imitate, the relevant human skills. Not that the heuristic approach has been ignored. On the contrary, heuristic problem-solving and the programming of "expert system" have been prominent computational methodologies in Artificial

Intelligence and Operations Research from the beginning. But in mathematical theorem proving, at least, the *purely logical* approach has far outpaced the heuristic approach.

The fact is that the latter has been severely hampered by a shortage of insights into mathematical cognition and ratiocination. Professor Alan Bundy's group at the University of Edinburgh has for some time now been patiently and insightfully seeking to remedy this shortage. Mateja Jamnik developed the ideas described in this book as a member of Bundy's group, and the book beautifully illustrates what the group has been doing.

What is novel about Mateja Jamnik's work is that she has found an explanation of at least part of the mystery of how humans are able to "see" the truth of certain mathematical propositions merely by contemplating appropriate diagrams and constructions. This ability to "see" is one of the really fundamental components of the human mathematical cognitive repertoire. As the late great mathematician G. H. Hardy put it: in the last analysis there is no such thing as "proof" – all a mathematician really does is *observe what is there.* To convince others of what he observes to be the case all he can do is *point* and say: *do you see?* Mateja Jamnik's program DIAMOND "sees", for example, as we do, that a 5×5 array of dots is also a nest of 5 "ells" containing respectively 1, 3, 5, 7, and 9 dots, and that there is nothing special about this special case. It and we can see that the case of 5 is but one instance of the *general* pattern whereby, *for any* n, an $n \times n$ array of dots is also a nest of n "ells" containing respectively $1, 3, 5, 7, \ldots, 2n-1$ dots. In this way it and we can directly see, as a kind of mathematical sense datum, the truth of the mathematical theorem that the number n^2 is the sum of the first n odd numbers. This act of seeing is analyzed in the program as a set of alternative decompositions of a given square array; we can see it either as a *row of columns*: or as a *column of rows*, or as a *nest of frames*, or as a *nest of ells*, or as an *array of subsquares*, and so on and on. Each of these perceptions is the same as seeing the truth of a corresponding mathematical theorem.

It is as though the program Deep Blue had been given some of the very same abilities to "see" the right chess move as Kasparov – literally to *see what he sees* when he looks at the board. But we don't know what it is that Kasparov sees.

Impressive though some of its achievements have been (such as that of EQP mentioned above), the *purely logical* approach to mathematical theorem proving is limited in scope and scientifically unedifying. The present book is an encouraging demonstration that its scope can be much widened, and its explanatory power expanded, by a fearless and patient exploration of the details of actual mathematical cognition. The logical and the heuristic approaches are beginning to come together in a most fruitful way. Mateja Jamnik is among the pioneers of a fresh new approach to an old problem.

I welcome the elegant and surprising insights of this book as ushering in a new generation of discoveries in the understanding of mathematical reasoning.

J. A. ROBINSON
Professor Emeritus, Syracuse University

Preface

This book started as my doctoral dissertation (Jamnik 1999) under the supervision of Alan Bundy and Ian Green at the University of Edinburgh, in what was then known as the Department of Artificial Intelligence, and had by the end of my PhD studies become the Division of Informatics.

This book is written for a mathematically minded audience. The intention is to present an exploration into the subset of the world of mathematics which can be solved with the use of pictures. Some familiarity with logic is required to understand the entire contents of the book. However, readers with little or no knowledge of logic should be able to safely omit parts of three particularly technical chapters: most of Chapter 8 and Appendix B, and parts of Chapter 4. Chapter 8 and Appendix B, in particular, are not essential to understand the general line of argument taken in the book. The specialist audience that the book is intended to attract is the automated reasoning community. The general audience that it is intended to attract is a community of scientists in artificial intelligence, computer scientists, mathematicians, philosophers, psychologists, cognitive scientists, teachers of mathematics and anybody interested in mathematical recreations.

I have attempted to make this book self-contained, and have included a comprehensive survey of other related work (see Chapter 2). In order to explain any possibly unfamiliar or unconventional use of terminology, I have included a Glossary of such terms at the end of the book.

Parts of the work described in this book have appeared in press in the past, in particular in Jamnik et al. 1999, Jamnik et al. 1998, Jamnik et al. 1997b and Jamnik et al. 1997a.

Acknowledgments

First and foremost, I would like to thank Alan Bundy for his invaluable advice, guidance and encouragement in my research. He has been an inspiration for me during these years.

I am indebted to Alan Robinson for not only being the first person to encourage me to publish this work as a book, but also for writing the foreword and giving me useful comments on a draft.

I would also like to thank Ian Green, Predrag Janičić, Nigel Shadbolt, Aaron Sloman and Keith Stenning for many useful discussions and comments on my work. Thanks to Manfred Kerber, Bob Lindsay and Toby Walsh for their comments on a draft of this book.

I especially thank Gavin Bierman for his love, understanding and encouragement throughout this time.

Lastly, to my family: *najlepša hvala moji družini za njihovo stalno ljubezen in podporo.*

This work was supported by a studentship from the Department of Artificial Intelligence at the University of Edinburgh, by a supplementary grant from the Slovenian Scientific Foundation, by a studentship from the British Overseas Research Scheme, and by EPSRC grant GR/M22031.

1

Introduction

$$n^2 = 1 + 3 + 5 + \cdots + (2n - 1)$$

— NICOMACHUS OF GERASA (*circa* A.D.100)
in NELSEN's *Proofs Without Words*

This book is about mathematical reasoning with diagrams. Human mathematicians often informally use diagrams when proving theorems. Diagrams seem to convey information which is easily understood by humans. For example, it requires only basic secondary school knowledge of mathematics to realize that the diagram above is a proof of a theorem about the *sum of odd natural numbers*. We call such proofs diagrammatic proofs. In this book we present an investigation into formalizing and mechanizing diagrammatic reasoning, and a concrete result of this investigation, a semi–automatic formal proof system, called DIAMOND (**Dia**grammatic Reas**on**ing and **D**eduction), which facilitates a user to prove theorems of arithmetic using diagrams.

1.1 Motivation

It is an interesting property of diagrams that helps us to "see" and understand so much just by looking at a simple diagram. Given some basic mathematical training and familiarity with spatial manipulations, we not only know what theorem the diagram represents, but we also understand the proof of the theorem represented by the diagram and believe it is correct.

Is it possible to simulate and formalize this sort of diagrammatic reasoning on machines? Or is it a kind of intuitive reasoning particular to humans that mere machines are incapable of? Roger Penrose claims that it is not possible to automate certain diagrammatic proofs.[1] We are taking his position as a challenge and are trying to capture the kind of diagrammatic reasoning that Penrose is talking about so that we will be able to emulate some simple examples of it on a computer. Our primary motivation is not to discover diagrammatic proofs, but to study them in order to understand them better and be able to formalize them.

The importance of diagrams in many domains of reasoning has been extensively discussed by Larkin and Simon (1987), who claim that "a diagram is (sometimes) worth *ten* thousand words". The advantage of a diagram is that it concisely stores information, explicitly represents the relations among the elements of the diagram, and it supports a lot of perceptual inferences that are very easy for humans. Diagrams have been extensively used in the history of mathematics to *aid informal mathematical reasoning*. The use of diagrams in explanations of theorems and proofs of geometry dates back at least to Ancient Greece, and the time of Aristotle and Euclid. Thus it is surprising perhaps that more recently, starting with the invention of formal axiomatic logic in the sense of Frege, Russell and Hilbert, diagrams have been denied a *formal* role in theorem proving. It is generally thought by logicians that diagrams have no accepted syntax nor semantic theory which would make them rigorous enough to be used in formal proofs. Hence, in the past century, only symbolic proofs of some logic have been considered to be *formal*, and proofs that use diagrams have been considered *informal*. Only very recently, in the last two decades, have there been efforts to fill this gap and investigate whether and how diagrams can be used in formal proofs (for instance, see Funt 1980, Sowa 1984, Kaufman 1991, Barker-Plummer and Bailin 1992, Barwise and Etchemendy 1994, Shin 1995, Hammer 1995, Stenning and Oberlander 1995).

Alongside the revival of research on formal aspects of using diagrams, investigations have also been carried out in other directions with different perspectives on the use of diagrams. These can be characterized into two groups of research perspectives, namely computational and cognitive perspectives.

From a computational perspective, Lindsay (1998) devises a compu-

[1] Roger Penrose presented his position in the lecture at the International Centre for Mathematical Sciences in Edinburgh, in celebration of the 50th anniversary of UNESCO on 8 November, 1995. His point of view is elaborated in Penrose 1994a.

tational model of human reasoning with diagrams, and claims that diagrams are sometimes more efficient for solving problems than some logical machinery. Glasgow and Papadias (1992) make a distinction between visual and spatial reasoning: visual reasoning is concerned about *what* a diagram looks like, whereas spatial reasoning deals more with *where* a diagram is located relative to other diagrams. Stenning and Oberlander (1995) introduce computational models for interpreting Euler's circles (Euler 1795). They also carry out a comparative analysis of the expressiveness of diagrammatic and symbolic representations in Stenning and Oberlander 1992. One of the aspects of the computational perspective is also the issue of knowledge representation. A lot of work on various kinds of representations has been carried out by Sloman and Hayes (see Sloman 1971, Hayes 1974, Sloman 1996). Related to this work and to Glasgow's work mentioned above is an unresolved debate on the characterization of diagrammatic (or graphical or visual) and symbolic (or sentential) representations (e.g., see Narayanan 1992, Olivier 1996, Blackwell 1997, Anderson 1997, Anderson et al. 2000, Anderson et al. 2001). Since there is no consensus on the definition of each type of representation, it seems that researchers adopt their own definitions suitable for their work.

From a cognitive perspective, Johnson-Laird (1983), and Hegarty and Just (1993) argue that humans, at least in some cases, use diagrams in their mental models of a situation. Mental imagery has been studied by Pylyshyn (1981), Pinker (1985) and Kosslyn (1993), amongst others. Pylyshyn is particularly critical of mental imagery and questions claims that humans use diagrams in cognition – we may think we do, he says, but there is no conclusive evidence that the brain uses diagrammatic representations; we may even be using symbolic logical representations (Pylyshyn 1973).

Our work contributes to research on formal and computational aspects of the use of diagrams, especially in automated reasoning systems. Automated reasoning systems have their roots back in the fifties when the first programs were written that could automatically prove simple theorems of propositional logic. As a result of growing interest in the research on automated reasoning we have today many sophisticated systems such as the theorem prover of Boyer and Moore (see Boyer and Moore 1990) or Isabelle (see Paulson 1989) in which one can prove complex theorems of mathematics.

However, during all these years, perhaps due to the influence of axiomatic logic, the majority of researchers have concentrated their efforts on improving the exact, rigorous and formal proof searching algorithms for a particular formal system of logic. In their efforts they have neglected the beauty and power of informal, intuitive reasoning of human mathematicians. There are exceptions including work by Gelernter (1963) and Bundy (1983). Bundy argued that in order to progress in computational logic, we need to go further and consider these informal aspects of human reasoning (Bundy 1983).

Our work supports this argument. We investigate informal human reasoning with diagrams and use it as an inspiration for formalizing diagrammatic reasoning so that it can be carried out on machines. We build a meta-theory in which diagrammatic proofs are formal. The issues which are addressed in this process include formality, informality and the rigor of diagrams in proofs. We hope to gain an insight into the understanding of at least a simple subset of diagrammatic proofs.

1.2 Aims

The concise storage of information, the intuitive representation of relations amongst elements of diagrams, and the support of perceptual inferences that humans seem to find easy to understand, are the characteristics of diagrams that we exploit in this book. We make the claim that most diagrams are "intuitive and easy to understand" informally, and support it only by anecdotal evidence from both, our own experience and that of some other people.[2] As mentioned before, there are at least two approaches to investigating diagrammatic reasoning corresponding to the perspectives on the diagrams research. One approach is to clarify and model the processes that are going on in humans when they use diagrams in mathematical reasoning – this can be described as a cognitive approach to investigating diagrams. Another approach is to design and implement a system which uses diagrammatic reasoning – this can be described as a formal and computational approach to investigating diagrams. In this book we take the second approach – our aim is to formalize diagrammatic reasoning and to show that diagrams can be used for proofs in a formal system.

Diagrams are concrete in nature. Unless we use *abstraction devices*[3] to represent the generality of a diagram, the diagram is a particular

[2]An experimental study into quantifying "intuitiveness" of diagrams and their use in mathematical proofs, and examining whether people find them "easier" to understand than symbolic logical proofs would be an interesting cognitive investigation.

[3]Note that in this book the word *abstraction* has two meanings due to a lack of two different appropriate words. First, an abstraction refers to some *abstraction device*, such as ellipsis (ellipsis is a term used for the "..." notation). Second, it refers to the *abstraction mechanism* which constructs a general proof from examples of a proof. The use of both meanings will always be clear from the context. Definitions of some terminology specific to this book, to which the reader is advised to pay special attention, can be found in Glossary on page 185.

instance of the general class to which it belongs – it is a typical representative instance for this classes. Abstraction devices are tools for representing the continuation of some pattern and are often used in objects to represent their generality. Examples of abstraction devices include ellipsis, or the summation of numbers sign $\sum$, or labelling of objects with variables. The use of abstraction devices in diagrams seems to be problematic, because it is difficult to keep track of them while manipulating a diagram. It is not clear if humans manipulate such abstraction devices or they reason with concrete objects and infer the generality in some other way. We aim to capture diagrammatic proofs which do not use abstraction devices on a computer. We use the concreteness property of diagrams and look into how theorems of mathematics can be expressed as diagrams for some concrete values, i.e., ground instantiations of a theorem.

The initial diagrams which represent (part of) a theorem are manipulated using some geometric operations which deconstruct diagrams in different ways, but preserve certain properties. For instance, if a diagram represents a natural number, then the collection of diagrams which is a result of applying some operation to the initial diagram represents the same natural number. This is true, because the operations are defined so that they preserve the natural number that the diagrams represent. The sequence of geometric operations on a diagram represents the *inference steps* of a diagrammatic proof. This is a novel approach to proving arithmetic theorems, which to the best of our knowledge, has not been undertaken before in other research on the automation of diagrammatic reasoning (see the overview of the past research in this field in Chapter 2). Rather than using symbolic formulae of some logic to prove a mathematical theorem, we use manipulations of diagrams. Our intuition is that the fact that the operations are visual seems to make them intuitively easier to understand and use for humans. No specialized knowledge of logic is required, just some familiarity with spatial manipulations. A concrete proof instance is called an *example-proof*, and consists of a sequence of operations applied to the concrete diagram. The set of all available diagrams and operations defines the proof search space.

Since manipulating abstraction devices to infer the generality of a diagram, or a theorem and its proof that the diagram conveys, can be problematic and can lead to ambiguous results, we need to find an alternative mechanism to capture a general proof of a theorem at hand. We do so by extracting a general pattern from several proof instances, and capture it in a recursive program, called a *schematic proof*. This recursive program allows us to construct a general diagrammatic proof for the universally quantified theorem at hand.

Finally, a general schematic proof which is inferred from the instances has to be shown to be correct. It seems that humans sometimes omit this step all together.[4] Human machinery for extracting a general argument is usually convincing enough to reassure them that the general argument is correct, e.g., consider the proof at the beginning of this chapter. In an automated reasoning system, we need to show the correctness of the induced general argument. This confirms that a diagrammatic schematic proof is indeed a correct formal proof of a theorem. We use the *constructive ω-rule*, an existing technique in logic (Sundholm 1983), to justify the step from schematic proofs to theoremhood. Baker et al (1992) investigated this rule in the domain of arithmetic theorems. The constructive ω-rule allows us to capture infinitary concepts in a finite way using the diagrams. In this book we aim to investigate the entire process of constructing examples, constructing a general proof, and showing that the general proof is correct. Together, all three stages constitute our formalization of *diagrammatic proofs*.

Having formalized the use of diagrams in proofs, it is no longer true that diagrammatic proofs can only be informal proofs. It is now interesting to investigate the relation between symbolic and diagrammatic proofs. Usually, theorems are symbolically proved with the use of inference steps which often do not convey an intuitive notion of truthfulness to humans in quite as easy way as diagrams do. The inference steps of a formal symbolic (as opposed to diagrammatic) proof are statements that follow the rules of some logic. The reason we trust that they are correct is that the logic has been previously proved to be sound. Following and applying the rules of such a logic guarantees that there is no mistake in the proof. We hope to have such a guarantee in our proof system, and moreover, to gain an insight into the intuitive understanding, the correctness and such properties of our diagrammatic proof. Ultimately, the entire process of diagrammatically proving theorems will illuminate the issues of formality, rigor, truthfulness and power of diagrammatic proofs, and perhaps more generally, of any sort of proof.

1.3 Some Original Contributions

There are three main contributions made by our work. First, our research introduces a novel approach to automated reasoning about mathematical theorems. There has been no work done on the automation of systems which use diagrams in such a direct way as our system Diamond, and the manipulations of diagrams lead to a correct proof of a theorem. All of the traditional formal rules of some logic which are expressed as

[4]Some anecdotal evidence will be given later in §4.5 and §4.6.

symbolic formulae, are completely replaced by geometric operations on diagrams. Thus, all the inference rules of DIAMOND are diagrammatic.

Second, the work presented in this book shows that diagrams *can* be used for *formal* proofs. Moreover, formal proofs are not just aided by diagrams, but can be constructed using only diagrams and operations on them. Although some people have claimed that diagrams can be given a rigorous function in reasoning and some people have disputed it, we are the first to show *how* it can be done in a particular subfield of mathematics so that formal diagrammatic rather than symbolic logical proofs can be generated. We formalize diagrammatic reasoning in a particular domain of mathematics (see Chapter 3), and implement a reasoning system DIAMOND which is capable of diagrammatically proving a number of theorems (Chapter 9). These proofs are guaranteed to be correct.

Finally, we show how the constructive ω-rule can be used to reason with particular instances of diagrams rather than with abstraction devices in general diagrams. We demonstrate how this technique can be used to capture general diagrammatic proofs (Chapter 4).

These three contributions are embodied in an implementation of a diagrammatic proof system called DIAMOND which automates diagrammatic reasoning and applies it to problem solving in mathematics. DIAMOND is a body of Standard ML code which interactively, via a graphical user interface, allows a user to construct diagrammatic proofs.

The construction of diagrammatic proofs in DIAMOND consists of three steps.

- The user interactively constructs example-proofs by choosing initial diagrams which represent the theorem (Chapter 5), and then applies diagrammatic operations (Chapter 6) to build these example-proofs.
- DIAMOND then automatically constructs a general pattern from these instances of proofs, and captures it in a recursive program, called a schematic proof. (Chapter 7)
- The final step is to check if the schematic proof is correct. DIAMOND automatically verifies a given schematic proof. (Chapter 8)

The main limitation of DIAMOND is that its expressiveness of diagrammatic rules is restricted. There are rules which cannot be expressed as manipulations of diagrams with the current repertoire. Indeed, there are theorems which consist of terms that cannot be expressed as diagrams. To overcome these weaknesses DIAMOND needs to be extended with some additional types of concrete diagrams and operations on them. Finally, DIAMOND is a proof checker, it is not a discoverer. The user of DIAMOND provides most of the intelligence by constructing example-proofs. Therefore, in order to enable DIAMOND to find proofs for itself, we could extend DIAMOND to a fully automated theorem prover which *discovers* diagrammatic proofs (Chapter 10) – this remains an interesting direction for future work.

There is a potential for the ideas we present in this book to be used for exploring human intuitive reasoning in a novel way. We think that humans find diagrammatic proofs easier to understand and more compelling than their symbolic logical counterparts. We have only anecdotal evidence to support our belief. However, some comparative psychological validity experimental study could be carried out. We propose that such a study could use DIAMOND to provide an architecture where the diagrammatic proofs can be constructed and explored in order to gain an insight into the understanding of the proof.

1.4 Layout of the Rest of This Book

Here is the organization and the layout of the rest of this book. It should give the readers an overall picture of the topics discussed in this book, and point them to a specific subject of interest.

In Chapter 2, *The History of Diagrammatic Systems*, we describe several other diagrammatic systems which have been implemented in the past. They all use diagrams for reasoning in some way: to store information, to reject false facts, to infer new facts, etc. We concentrate in more detail on Gelernter's Geometry Machine, Koedinger and Anderson's DC, Barker-Plummer and Bailin's Grover, Barwise and Etchemendy's Hyperproof, Lindsay's Archimedes, Furnas' Bitpict, and Anderson and McCartney's IDR, because these seem to be closest to our work with respect to the use of diagrams for problem solving.

In Chapter 3, *Diagrammatic Theorems and the Problem Domain*, we present some examples of theorems which can be represented and proved in a diagrammatic way. Diagrams are often perceived as an informal rather than formal aid to reasoning, so we discuss their use in proofs, and the general issues about the formal and informal role of diagrams in proofs. We then present some examples of theorems that can be proved diagrammatically by showing the diagrams and the manipulations on them. Based on these examples, a taxonomy of diagrammatic proofs is introduced. Another factor which is considered in our choice of the problem domain is the use of abstraction devices (e.g., ellipsis) in diagrams. Finally, the taxonomy helps us choose the domain of problems that we subsequently concentrate on in this book.

In Chapter 4, *The Constructive ω-rule and Schematic Proofs*, we give a way of capturing diagrammatic proofs without the need to resort to diagrams containing abstraction devices. The mathematical basis for capturing the generality of the proof is in the use of the constructive ω-rule in schematic proofs, which is explained in detail.

In Chapter 5, *Designing a Diagrammatic Reasoning System*, we describe the DIAMOND system which is an embodiment of the ideas presented in this book. DIAMOND is a diagrammatic proof checker, which interactively proves theorems of mathematics by applying geometric operations to diagrams. In this chapter some of the design issues for the implementation of this proof system are discussed. These include: the architecture of DIAMOND, the basic notion of a diagrammatic proof, the construction of example-proofs, the representation of diagrams, and DIAMOND's graphical interface.

In Chapter 6, *Diagrammatic Operations*, we present the geometric operations, which are available in DIAMOND. These operations capture the inference steps of a diagrammatic proof. We define them here and give some examples.

In Chapter 7, *The Construction of Schematic Proofs*, the notion of a diagrammatic proof is presented. A diagrammatic proof is captured in a recursive program, referred to as a schematic proof. When a schematic proof is run, it generates a proof of $P(n)$ for each input value of number n. In this chapter we describe how general schematic proofs are automatically constructed from example-proofs, and how they are formalized in DIAMOND.

In Chapter 8, *The Verification of Schematic Proofs*, we present a method which enables us to prove the correctness of schematic proofs for *particular* theorems. The mechanism for construction of a schematic proof is an inductive inference algorithm. It is a machine's attempt to make an "intelligent" guess of what the general proof is. This "guess" needs to be verified and shown to be correct. In this chapter we define a way of carrying out the verification, in particular, we devise a theory of diagrams where we can check the correctness of a schematic proof.

In Chapter 9, DIAMOND *in Action*, we present a running example of the construction, formalization and verification of a diagrammatic proof in DIAMOND. We also comment on the general results in DIAMOND such as the range and depth of theorems it can interactively prove, and the limitations of DIAMOND.

In Chapter 10, *Complete Automation*, we propose some possible future directions for the work discussed in this book. In particular, we give an indication of how to make DIAMOND a completely automated theorem prover capable of discovering diagrammatic proofs. Finally, we make some concluding remarks.

In Appendix A, *More Examples of Diagrammatic Theorems*, we give additional examples of theorems and their diagrammatic proofs, which are analyzed to motivate the taxonomy of diagrammatic theorems used to choose the problem domain discussed in this book.

In Appendix B, *The ω-Rule*, we define and motivate the use of the ω-rule in logic. The problems with its use in automation lead us to the use of its constructive version (explained in Chapter 4).

In Glossary we give some definitions of technical terms used in this book that might prove useful. Notice that in the literature, the terms induction, abstraction and generalization are often used interchangeably for the same concept. We have three different notions for these terms, and hence define them here precisely. We urge the reader to pay particular attention to the use of these terms.

2

The History of Diagrammatic Systems

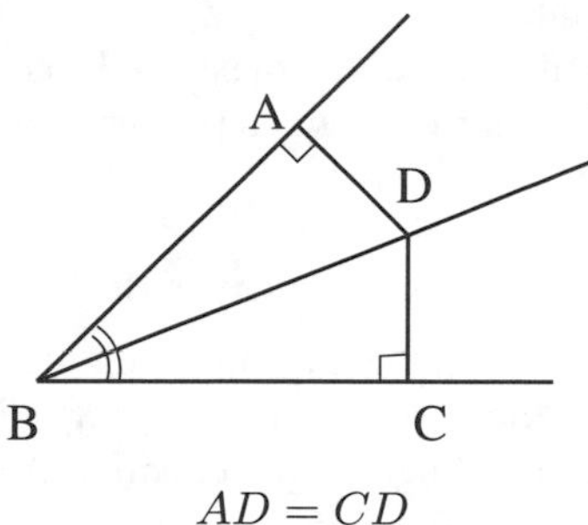

$AD = CD$

— H. GELERNTER
Realization of a Geometry–Theorem Proving Machine

This chapter is a survey of previously implemented systems which mechanize the use of diagrams for reasoning. The systems which are perhaps closest to our work on the use of diagrams for problem solving are Gelernter's Geometry Machine, Koedinger and Anderson's DC, Barker-Plummer and Bailin's Grover, Barwise and Etchemendy's Hyperproof, Lindsay's Archimedes, Furnas' Bitpict, and Anderson and McCartney's IDR. We also mention the work by Shin and Hammer which tries to establish the soundness of diagrammatic reasoning. Finally, we briefly mention other interesting, but perhaps less related research.

Roughly, diagrammatic reasoning systems are computer programs which use a diagram to aid the search for the solution of some problem. The first such program was Gelernter's Geometry Machine. Others share much with Gelernter's Geometry Machine, e.g., the problem domain of Euclidean plane geometry.

We distinguish between visual and diagrammatic representations. A visual representation is a visual display of a diagram on some medium so that it can be seen by the user. A diagrammatic representation describes a diagram in some way which depicts its visual characteristics. For example, Cartesian coordinates describe the elements of a diagram by indicating their position in the coordinate system. A diagrammatic representation does not necessarily have to be presented visually so that the user can see it, e.g., visualize it on a computer screen. Instead, some non-visual representation may be used. For example, a diagram may be described using some predicates for relations among its elements. In most cases, diagrams are represented by Cartesian coordinates, in some cases by the bitmap or raster matrix, and in some cases they are in fact visual (e.g., the user interface allows the display of a visual image of the diagram). All of the above mentioned systems use representations that are diagrammatic, however, the representations vary in the degree to which they are visual.[5]

The systems presented here are described according to their architecture and their main features, with particular focus on their use of diagrams.

2.1 Gelernter's Geometry Machine

The first implemented system which used diagrams for reasoning was Gelernter's Geometry Machine (Gelernter 1963). The novelty of Gelernter's work was its use of a diagram as a model of the goal to be proved to control the search for a proof of a theorem. In the beginning of this chapter we showed an example of a theorem and a diagram which the Geometry Machine used to prove the theorem.

The Geometry Machine operated on statements expressed as strings of characters in a formal logical system.[6] The problem is a statement, and the solution, i.e., the proof, is a sequence of statements. A proof of a theorem starts from some axiom that the system chooses. Then it continues inferring further theorems based on the existing axioms or on other theorems. The final statement of the proof is the problem itself.

Working from the axioms in a complete theory ensures that the sequence under consideration as a proof indeed terminates in the required theorem. However, the problem-solving tree still has a high degree of branching. To prune the search tree, the Geometry Machine uses heuris-

[5] Related to the discussion about the difference between visual and diagrammatic representations is Glasgow's work (Glasgow and Papadias 1992) where she distinguishes between visual (what the diagrams look like) and spatial (where the diagrams are with respect to other diagrams) representations.

[6] The reader is referred to Gilmore's rational reconstruction of Gelernter's Geometry Machine for a more formal definition of its logical theory (Gilmore 1970).

tic properties of a diagram to reject false subgoals. This means that the subgoals are tested against measurements of a coordinate diagram, and if the subgoal is false in the diagram, then it is rejected.

The Geometry Machine consists of three components:

Syntax/Logic: (also called a syntax computer) it manipulates the formal system by generating strings of hypotheses (e.g., premises, subgoals).

Model/Semantics: (also called a diagram computer) the theorem to be proved is represented by a coordinate system. Also, it contains a series of qualitative descriptions of the diagram.

Search Control: (also called a heuristic computer) it is the main component of the system. It compares sequences of strings generated by the syntax component and their interpretation in the diagram. The search control component rejects subgoals not supported by the diagram. Furthermore, it recognizes the symmetries amongst classes of strings and reduces the search space accordingly.

The flow of control in the Geometry Machine is such that it allows the syntax component to communicate with the model component and vice versa only through the search control component (see Figure 1 which was adapted from Gelernter 1963).

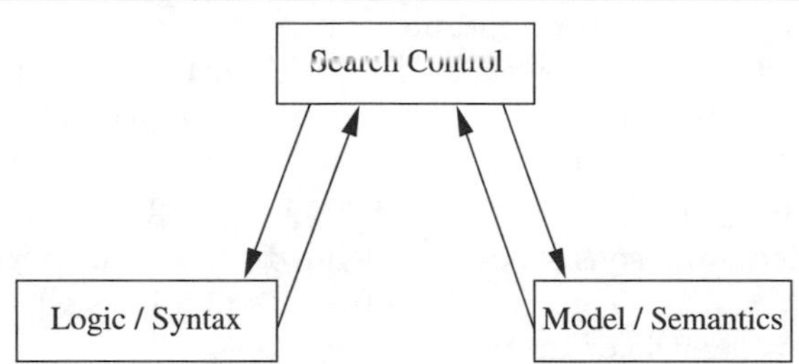

FIGURE 1 The architecture of the Geometry Machine.

It is important to note that the system does not generate its own diagram. Rather, the diagram is supplied by the user. The diagram is supplied to the Geometry Machine in the form of a list of coordinates for points named in the theorem. A second list, also supplied by the user, specifies points joined by segments.

The diagram has two roles. Its *negative* role is to reject hypotheses (subgoals) proposed by the search control component that are not true in the diagram. In this way the search space is pruned. The *positive* role of the diagram is to shorten the inference paths by assuming various facts that are obvious in the diagram as true, i.e., it verifies the correctness of simple goals by checking them in the diagram (e.g., a certain point lies between two others).

In summary, Gelernter's Geometry Machine is a theorem prover guided by a user-supplied model in the form of a diagram. It uses a diagrams to prune the search for a symbolic proof.

2.2 Koedinger and Anderson's DC

Koedinger and Anderson (1990) implemented a geometry problem solver called the Diagram Configuration (DC) model. An interesting characteristic of this system is that the authors based the configuration of the model of the system entirely on empirical data from testing how human experts solve geometry problems. Thus, supported by their empirical evidence, they claim that DC reasons the way humans do.

The key feature of the system is that its data is organized in perceptual chunks, called diagram configurations. These are analogous to key features of diagrams that humans recognize when they inspect a diagram. During the process of generating a solution path, DC infers the key steps first, and ignores along the way the less important features of the input diagram, i.e., the less important inference steps.

The Diagram Configuration model (DC) consists of:

Diagram Configuration Schemas: these are major knowledge structures of DC. They are associated with elementary or more complex geometric structures in the form of clusters of geometry facts (e.g., congruent-triangles-shared-side scheme, perpendicular-adjacent-angles scheme). A scheme consists of the following parts:

- **Configuration:** consists of a predefined geometric image, i.e., a diagram. It is a configuration of points and lines which is part of the geometric diagram.[7]
- **Whole-statement:** is a geometry statement referring to the whole of the configuration (e.g., $\triangle XYZ \cong \triangle XZW$).
- **Part-statements:** are geometry statements referring to the relationships among the parts of the diagram (e.g., $\angle Y = \angle Z$).
- **Ways-to-prove:** lists subsets of part-statements that are sufficient to prove the whole-statement and hence all of the part-statements.

DC's Processing Components: DC consists of three major processing stages:

- **Diagram Parsing:** it recognizes configurations in the diagram in-

[7] Note that this is a diagram of the schema and *not* an input diagram to DC.

put by the user and instantiates their corresponding schemas. The recognition is done on two levels: low-level simple object recognition and high-level plausible configuration hypothesizing.

Statement Encoding: it deciphers the meaning of the given and goal statements, and represents them as part-statements which are tagged either "known" or "desired".

Schema Search: using forward and backward inferences, schemas that are possibly true in the problem are iteratively identified (i.e., the system searches through possible schemas until the link between the given and a goal statement is found).

Note that a whole-statement can be viewed as a conjecture of the schema, and ways-to-prove are hypotheses which are sufficient to prove the conjecture provided that the hypotheses are proved as well.

The main idea of DC is that it uses schemas instead of statements of geometry to plan the search for solution to a problem. In the first stage, the input diagram is parsed and the possible schemas are instantiated. This is done by inspecting the elements of the input diagram and identifying the schemas that are related to particular features of the input diagram (for example, if the input diagram contains a right angle triangle then the schema for right angle triangles is instantiated). Hence, the input diagram triggers the identification of several schemas. However, a configuration of the schema might have other features that are not identified by the parsing of the input diagram. DC adds such schemas to the solution space as well. Hence, establishing one schema may enable establishing another. No problem solving search is done at this stage, however, the biggest part of the work of the system is done by restricting the solution space by input diagram parsing. Figure 2 shows a problem definition and the solution space of the problem after the diagram parsing and the instantiation of schemas (taken from Koedinger and Anderson 1990). The boxes show the schemas that have been recognized and the lines connect schemas to their part-statements.

After diagram parsing, the given/goal statements of the problem definition are encoded by tagging them as "known" (or "desired") if they are already part-statements or whole-statements of a certain schema. Finally, to find the solution the system searches for a path from the given statements to the goal statements. Note that the constraints which are listed in the ways-to-prove component of the schema have to be met when searching for the solution path. There may be several solution paths.

In summary, Koedinger and Anderson's DC system controls the search

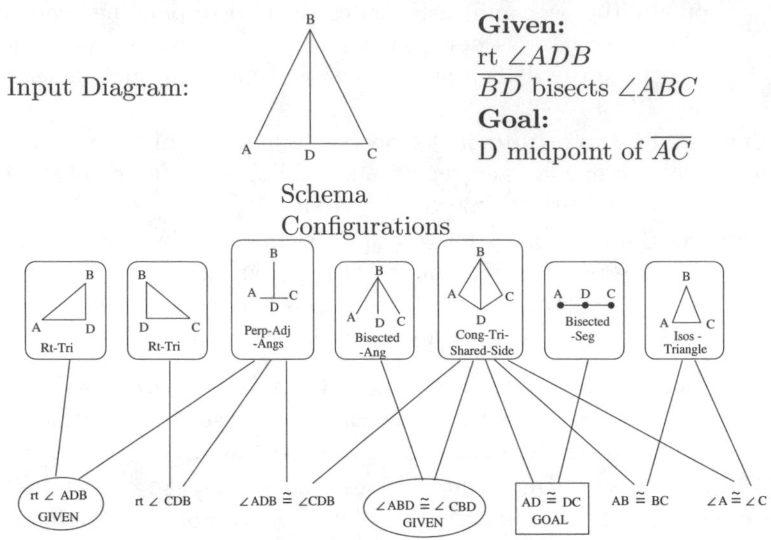

FIGURE 2 DC's problem definition and solution space.

for a solution of a problem by organizing the proof search space into smaller spaces which deal with specialized concepts, i.e., schemas. These, when identified to be related to a problem, restrict the set of rules that DC can apply to find a proof. DC's schemas can be thought of as derived rules of inference which are identified by the diagram and can be applied in the proof. Like the Geometry Machine, DC too uses a diagram to search for a symbolic proof.

2.3 Barker-Plummer and Bailin's "&"/Grover

"&"/Grover, developed by Barker-Plummer and Bailin (1992) is an automated reasoning system which uses information from a diagram to guide proof search.

The architecture of "&"/Grover system consists of the "&" automated theorem prover, based on the sequent calculus for Zermelo Frankel Set Theory,[8] and Grover which is the diagram interpreting component of the system. Grover passes the crucial information to prove the theorem from the inspected diagram to the "&" theorem prover. In the scope of this book we are mainly interested in the Grover diagrammatic reason-

[8]See Bailin and Barker-Plummer 1993 for more information on Zermelo Frankel Set Theory.

ing component. The architecture of Grover is shown in Figure 3,[9] and

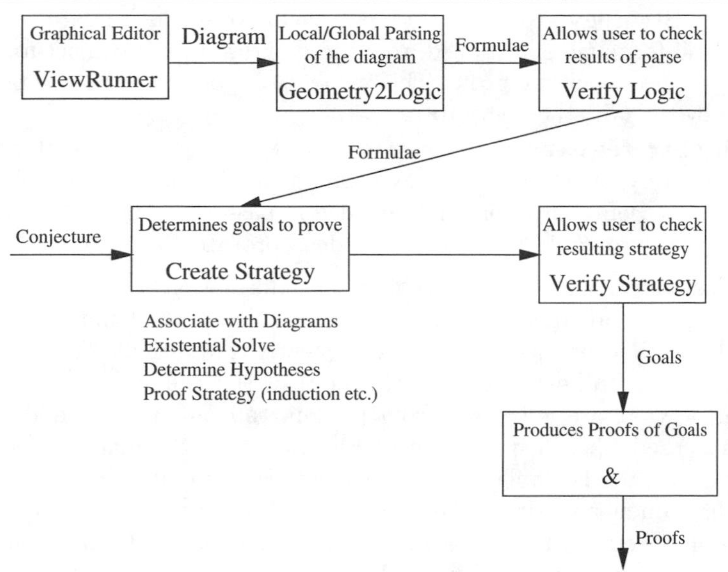

FIGURE 3 The architecture of Grover.

consists of the following components:

ViewRunner: is the graphical editor tool, and Grover's interface. It enables users to draw a diagram consisting of fairly elementary components. The diagram is saved as an abstract description of the geometry of the diagram (i.e., describing arcs, circles, arrows, dots, etc.).

Geometry2Logic: is an expert system component. It parses the abstract description of the diagram and translates the logical content of the diagram into formulae expressed in the "&" language. It works in a bottom-up fashion. This means that it first analyses the objects of the diagram, then relationships between these objects and finally, the collection of atomic formulae to determine more complex formulae.

Verify Logic: is an inspection tool allowing the user to examine and modify the logical content (i.e., logical formulae description) of the

[9]The figure of the architecture is taken from the unpublished paper by D. Barker-Plummer, S.C. Bailin, and S.M. Ehrlichman, entitled "Diagrams and Mathematics" from 1995, supplied to me by D. Barker-Plummer.

input diagram. This description is derived by the Geometry2Logic component from the graphical representation of the diagram.

Create Strategy: constructs a sequence of goals which relate the logical formulae determined from the diagram to the conjecture that the user wants to prove. This sequence of goals can then be proved by the "&" theorem prover.

Verify Strategy: allows the user to inspect the sequence of goals generated by the Create Strategy component. If it is decided that they are acceptable, then the sequence is passed to the "&" theorem prover to verify that they are indeed provable.

The main idea in Grover is that the information is extracted from the diagram and translated into logical formulae in the language of "&" which are then proved by "&". Subsequently, the formulae are used as additional hypotheses to the main proof of the conjecture. Thus, the formulae that are extracted from the diagram are in fact additional lemmas used when searching for a proof in "&" of the main conjecture.

"&"/Grover is similar to the Geometry Machine in that it also uses the diagram as a model of the goal which is to be proved. Moreover, the diagram specifies the subgoals themselves. Therefore, it constrains the high-level structure of the proof. Also, it specifies the ordering in which the subgoals are applied. In order to prevent a high degree of branching of the proof search tree, Grover considers only subgoals that are known to be true in the diagram, and in this way prunes the proof search space in "&".

2.4 Barwise and Etchemendy's Hyperproof

Hyperproof by Barwise and Etchemendy (1991) is an educational tool for teaching logical reasoning, and in particular first-order logic. Its domain of reasoning is a blocks world. The system uses a symbolic representation of first-order logic, as well as a diagrammatic representation to describe situations in the blocks world. The user learns how to construct proofs of both consequence[10] and non-consequence[11], proofs of consistency and inconsistency, and independence[12] proofs. Hyperproof automatically checks the logical validity of each type of proof.

A proof in Hyperproof starts with a blocks world situation described in a diagrammatic form using a graphical display. This is the initial infor-

[10] A proof of consequence is an argument which establishes a proposition from a set of givens.

[11] A proof of non-consequence demonstrates from the set of givens that a proposition may not hold.

[12] An independence proof shows that a proposition cannot be proved on the basis of a set of givens.

mation for the proof. In addition some sentences of first-order logic might be given using the symbolic representation. All of the initial information is called *given information.* The aim is to show that some conjecture about the given information is a consequence or a non-consequence of the given information. Such a conjecture is normally represented using a symbolic representation.

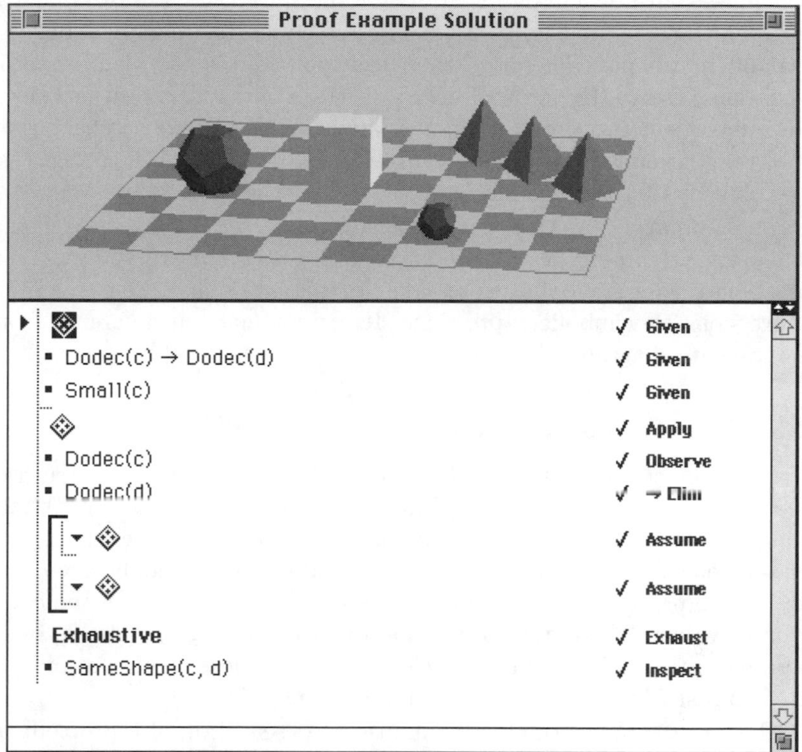

FIGURE 4 Hyperproof's proof.

Figure 4 gives an example of the type of reasoning that Hyperproof is designed for. The picture in the upper part of Hyperproof's screen is the initial information given. The aim is to determine whether block c and block d are of the same shape: $SameShape(c, d)$. This is indeed a consequence of the given information. The given information consists of two sentences: $Dodec(c) \rightarrow Dodec(d)$ and $Small(c)$. The diagrammatic situation in the upper part of the Hyperproof screen can be accessed from the bottom symbolic part of the screen by clicking on the diamonds ◈ which form some of the lines of the symbolic proof.

The first step in the proof applies the second piece of symbolic information to the diagrammatic situation (see Figure 4). Namely, it identifies the only small block in the situation and labels it with c. The user then observes that c in indeed a dodecahedron. Using the first piece of given symbolic information, i.e., $\mathit{Dodec}(c) \rightarrow \mathit{Dodec}(d)$, the user then concludes that d is a dodecahedron as well. Since there are two dodecahedra in the diagrams, there are two possible ways of assigning label d to two different blocks. Therefore, there are two possible new diagrammatic situations. Hence, Hyperproof describes these two possible situations in the subsequent two proof steps, and they can be viewed by clicking on the two diamonds ◈ in Figure 4. In each of the two situations the user can observe that the two blocks c and d have the same shape.

In summary, Hyperproof is a heterogeneous reasoning system since it can make three types of inferences: from diagrams to symbolic expressions, from symbolic expressions to diagrams and from symbolic expressions to symbolic expressions. It cannot make an inference from diagrams to diagrams.

2.5 Lindsay's Archimedes

Lindsay's Archimedes system (Lindsay 1998) demonstrates the truth of theorems of Euclidean geometry, e.g., *Pythagoras' theorem* (see §3.2.2). Diagrams are represented as a combination of points and segments in a two-dimensional coordinate system, plus additional symbolic sentences describing names, facts and constraints about the diagram in the coordinate system. The diagrams are manipulated through rotations, overlapping, symmetries, adding new elements, etc. These manipulations are used to justify the truth of a posed statement.

The user supplies the diagrammatic representation of a problem by specifying the diagrams that represent the theorem. The user also supplies the operations that need to be applied to diagrams in the process of constructing a justification of this statement. Archimedes uses its representation to infer new facts that emerge as consequences of applying these operations to the diagram. It can then relate these new properties and relations to symbolic statements. Archimedes can observe new emerging diagrams described by the existing elements in the diagram, e.g., a new diagram is formed from the existing points. But it cannot observe exhaustively all new diagrams formed from elements that are created as a consequence of operations. The reason is that this is computationally too expensive, hence Archimedes observes simple facts, but it only notices the complex ones when instructed to look for them. For example, when a segment *ab* is constructed because of some operation, and it crosses with another segment *cd*, then a new point is formed. Archimedes can observe that this new point was created and it can name it, say *e*. It can also detect that two new segments were formed, namely *ea* and *eb*. However, it cannot observe (unless instructed to do so by the user) that the segments *ec* and *ed* were also created. Archimedes also cannot observe unspecified emerging relations, e.g., two elements become congruent by construction. However, the user can guide the system explicitly to observe these emerging features in the diagram.

Since a demonstration of the truth of a theorem is made on a particular diagram, some generalization is required in order to conclude that the statement is true in all cases. Archimedes does not do any generalization of a demonstration of a statement. However, it does help the user of the system by performing a simulation of the demonstration. This simulation makes the generality of some operations and relations explicit, and hence more easily understood. In this sense, Archimedes is a theorem (or demonstration) checker.

Archimedes is related to our work in that it uses diagrammatic operations rather than symbolic expression manipulations as a major source of inference about the posed theorem. This makes our and Lindsay's work differ significantly from almost all prior work on diagrammatic reasoning. While the work on Archimedes is motivated by some cognitive issues (e.g., Archimedes could be used as a model of how humans sometimes infer the truthfulness of statements), and we share some of these motivations, we are also interested to show that such demonstrations are general and lead to correct formal proofs of theorems.

2.6 Furnas' Bitpict

Furnas' Bitpict system (Furnas 1990, 1992) reasons only with diagrammatic inference rules. The domain of problems is from qualitative physics. The primitive notions in Bitpict are bitmaps, i.e., regular grids of picture elements (pixels) that are either black or white. A set of bitmaps is specified by bitpicts, i.e., given a bitpict all bitmaps which contain this bitpict can be selected. A bitpict is a specific configuration of pixels. An inference rule in Bitpict consists of a pair of bitpicts. The first one is used to select bitmaps and specific locations in them where this bitpict occurs. The second one is used to rewrite with the second bitpict the location in a bitmap that was selected by the first bitpict. The top part of Figure 5 (taken from Furnas 1992) shows a simple diagrammatic rule in Bitpict consisting of a pair of bitpicts which rewrite a 3×3 grid

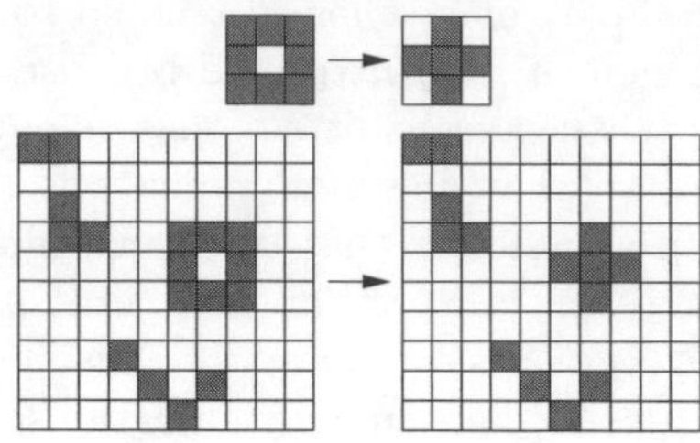

FIGURE 5 Bitpict's inferencing.

containing a circle into a 3×3 grid containing a cross. The bottom part of Figure 5 shows how this rule is used to rewrite a bitmap. Bitpict is a simple reasoning system based on the standard rule production system, where the usual symbolic rewrite rules are replaced by the so-called bitpict diagrammatic rewrite rules.

Examples of problems that Bitpict solves include counting the disconnected components in a tangled forest of bifurcating trees, playing tic-tac-toe, solving simple wheel-and-pulley rotation direction problems and running PacMan.

Bitpict is interesting in relation to our work in that it is one of the few systems that reasons with only diagrammatic inference rules. Moreover, its diagram representation, although inefficient for processing large diagrams, it depicts only spatial structures and is hence entirely diagrammatic.

2.7 Anderson and McCartney's IDR

Anderson and McCartney's Inter-diagrammatic Reasoning system (IDR) (Anderson and McCartney 1995) is similar to Bitpict in that it also uses a pixel representation of a diagram. Unlike Bitpict, the pixels in an IDR's bitmap can be of varying degrees of grey, with white and black as two opposing extremes. Anderson and McCartney distinguish between intra-diagrammatic reasoning (reasoning with a single diagram) and inter-diagrammatic reasoning (reasoning with related groups of diagrams). They define some general inter-diagrammatic operators that are useful for reasoning in a number of domains. These operators are defined over pixel representations of diagrams. Examples of such operators include negation *NOT* which turns all black pixels into white and leaves white pixels white; *AND* which returns a minimum of each pair of corresponding pixels in two diagrams, where the minimum is defined as the pixel whose value is closest to white. Figure 6 shows how two diagrams are combined using these operators in a game of battleship (Anderson and McCartney 1995).

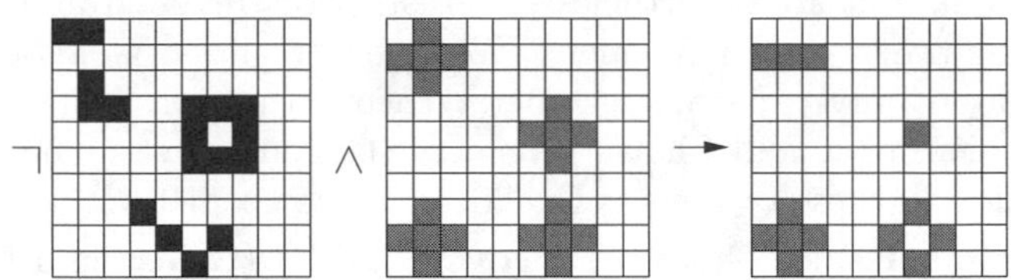

FIGURE 6 IDR's inferencing.

IDR was applied to reason in various domains including for developing heuristics for a battleship game, for inferring the correct fingering of a sequence of guitar chord diagrams and for inferring the quality of precipitation in a suite of cartograms. For details the reader is referred to the cited literature.

IDR is related to our work in that it operates on entirely diagrammatic representations, namely the pixel grids. Hence, similarly to our aims, it reasons with entirely diagrammatic inference rules.

2.8 Logic of Diagrams

Another related area of work formalizes the logic of diagrams. Similarly to our work, it aims to show that reasoning with diagrams can be sound and formal. Typical representatives are Shin (1991, 1995) and Hammer (1995).

Shin showed that two particular systems of Venn diagrams (Shin 1991, 1995) have a formal syntax, semantics and model theory. Hence, she showed that diagrams in a purely diagrammatic reasoning system can be used as formal tools.

Hammer extended Shin's work by allowing also symbolic inferences, rather than just diagrammatic ones. Hence, similarly to Hyperproof, Hammer's system is also heterogeneous: like Hyperproof, it can make inferences between symbolic statements and diagrams, diagrams and symbolic statements, and between symbolic statements and symbolic statements. In addition, like Bitpict, it can also make inferences between diagrams and diagrams. Hammer showed the formal syntax and semantics, and that the diagrammatic transformations are sound and complete (Hammer 1995).

2.9 Other Related Systems

Besides the diagrammatic reasoning systems presented so far, there exist several others. They are perhaps less related to the system described in this book, but are nevertheless interesting in the diagrammatic features that they use or implement. We briefly mention Goldstein's Basic Theorem Prover, Nevins' geometry theorem prover, McDougal and Hammond's Polya, and Chou's geometry theorem prover. The problem domain for all these systems is Euclidean plane geometry.

Goldstein's Basic Theorem Prover (BTP) (Goldstein 1973) extends Gelernter's Geometry Machine. BTP solves problems from a small part of plane Euclidean geometry. The input to BTP is a diagram, which is represented in the form of Cartesian coordinates of the points and a list of connections between the points, the hypotheses and the objective, i.e., the goal. This diagram is parsed and used to reject goals that are false in the diagram.

Nevins' Geometry Theorem Prover (Nevins 1975) uses a forward reasoning strategy. He claims this is the way humans think – although this is by no means resolved within psychology. Certain features of the diagram cue the inference steps, which are made using a number of *paradigms*. The paradigms are guided by the diagram and can make multiple conclusions. They are capable of making inferences that require multiple steps. In many ways Koedinger and Anderson's DC (see §2.2) system extends the Nevins model. However, Nevins' system does not visualize the diagrammatic model, nor does it use numerical information from the diagram.

McDougal and Hammond's Polya (McDougal and Hammond 1993) is a geometry theorem-prover. Its input is a list of givens, a goal and a diagram. Its output is a proof which is arrived at after a series of interpretations of plans for visual search and plans for writing proofs. The diagram is described in terms of Cartesian coordinates, marks for segments and marks for angles. For more information see McDougal 1993 and McDougal and Hammond 1995.

Chou's Geometric Theorem Prover Chou's geometric prover is one of the most powerful theorem provers of geometry (Chou 1988). Diagrams are used to infer some conditions in the proof, but the symbolic inference methods do not make much use of intuitive spatial properties of diagrams. Hence, the theorem prover constructs symbolic, rather than diagrammatic proofs.

In the domain of qualitative physics, the following systems are of interest: Funt's WHISPER (Funt 1980), and Iwasaki, Tessler and Law's REDRAW (Iwasaki et al. 1995). For more information the reader is referred to the cited literature.

There are several significant collections of papers on the work done in the area of diagrammatic reasoning – see, e.g., Narayanan 1992, Chandrasekaran et al. 1995, Anderson 1997, Meyer 1998, *Journal of Logic, Language and Information* (1999, volume 8, number 3), Anderson et al. 2000 and Anderson et al. 2001.

2.10 Summary

In this chapter we surveyed the history of diagrammatic reasoning systems. Most of the systems implemented in the past have Euclidean plane geometry as their problem domain. Gelernter's Geometry Machine was the first system which used a diagram to aid the search for the proof of a theorem. Similarly to the Geometry Machine, Grover uses the diagram to prune the search for the essentially symbolic proof. The Diagram Configuration model consists of derived rules about geometrical facts which are used to construct the search space if the problem is related to the diagram used in the rule. In Hyperproof the user can construct proofs by using first-order predicate logic rules *and also* the diagrammatic rules derived from the diagram situations in a blocks world. However, unlike in the other systems mentioned so far, the diagrammatic inference steps from diagrams to symbolic expressions and vice versa can be made. Apart from Hyperproof, some of the other systems which are related to our approach in that they use geometric manipulations of diagrams for inferencing include Archimedes, Bitpict and IDR.

One of our aims is to show that diagrammatic reasoning can be formalized. Some of the related research in this direction, i.e., that shows the soundness of diagrammatic reasoning, includes the work of Shin and that of Hammer on establishing the logics of diagrams. While some of our motivations are shared, their systems, unlike in our work, construct essentially *symbolic proofs* of theorems.

3

Diagrammatic Theorems and the Problem Domain

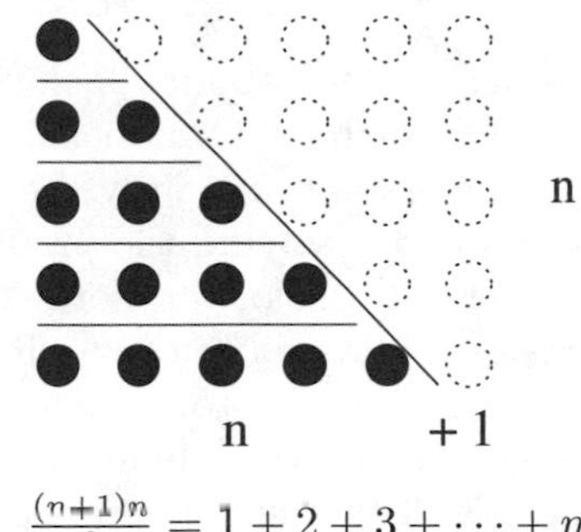

$\frac{(n+1)n}{2} = 1 + 2 + 3 + \cdots + n$

— "THE ANCIENT GREEKS" (as cited by Martin Gardner)
in NELSEN'S *Proofs Without Words*

One of the aims of this book is to show that proofs which use diagrams and manipulations of diagrams, rather than symbolic formulae of some logic, can be automated and emulated on a machine. Before mechanizing such diagrammatic proofs the class of theorems which lend themselves to diagrammatic representation needs to be identified. Once we know what type of theorems can be represented diagrammatically, and can be manipulated via some diagrammatic operations, we devise a taxonomy which enables us to choose the domain of problems on which we focus.

In this chapter we present some examples of theorems which can be represented and proved in a diagrammatic way. Diagrams are often perceived as an informal rather than formal aid to reasoning, so we discuss in §3.1 their use and the general issues about the formal and informal role of diagrams in proofs. In §3.2 we present some examples of theorems that can be proved diagrammatically by showing the diagrams and the manipulations on them. Based on these examples, a taxonomy of diagrammatic proofs is introduced in §3.3. Another factor which is considered in our choice of the problem domain is the use of abstraction devices (e.g., ellipsis) in diagrams, which is discussed in §3.4. Finally, in §3.5, the taxonomy helps us choose the domain of problems that we subsequently concentrate on in this book.

3.1 Diagrams and Proofs

> *"There is no more effective aid in understanding certain algebraic identities than a good diagram. One should, of course, know how to manipulate algebraic symbols to obtain proofs, but in many cases a dull proof can be supplemented by a geometric analogue so simple and beautiful that the truth of a theorem is almost seen at a single glance."* Gardner 1986

This is Martin Gardner's quote about "look-see" proofs. Gardner writes about diagrams that guide human mathematical thought, and enable a mathematician to understand instantly the problem represented by the diagram, *how* to go about solving the problem and *why* the solution is correct. In the everyday use of the word, "seeing" often means understanding. Diagrams as objects which convey information in a visual way often seem to be more easily understood than other representations, such as symbolic formulae. The debate about the formal and informal use of diagrams is a long standing one (Larkin and Simon 1987).

Diagrams were used to solve problems as far back as Ancient Greece. In those times there were two modes of representation that coexisted, but only rarely mixed. They were the Aristotelean logic, or what we now call symbolic reasoning, and Euclidean geometry which used diagrams plus algebra for inferencing. It was Descartes who brought the two modes of reasoning together, and showed that symbolic and diagrammatic reasoning can complement each other in solving problems. Descartes showed this by the invention of analytic geometry. However, at the turn of this century symbolic representation took over as the only rigorous mode of reasoning. The founders of modern logic, Frege, Russell and Hilbert, advocated that all arithmetic concepts be defined in logical terms, and all arithmetic knowledge be expressed and derived from the axioms and definitions of the logic. Reasoning was considered to be rigorous only if it was expressed in the formal language of some logic. The diagrammatic representation became neglected, not only due to the power that logic provided, but also due to some carelessly constructed diagrams, whose use turned out to be faulty (see Maxwell 1959 and Dubnov 1963). Diagrams lost their legitimate role in formal proofs. They were not thought to be rigorous and formal enough for use in proofs.

However, in the last twenty years, researchers from various fields, such as cognitive science, artificial intelligence, computer science, physics, and mathematics returned to the use of diagrams and tried to re-establish a formal role for diagrams in proofs. Some of the work has already been mentioned in Chapter 2, but let us also mention Shin's rigorous analysis of Venn diagrams as a formal system (Shin 1991, 1995), Sowa's work on Pierce's existential graphs in Sowa 1984, and the use of diagrams in category theory (MacLane 1971). It seems that the neglect of the formal use of diagrams in proofs has motivated many researchers to explore whether diagrams can indeed be part of a formal system, and whether the formal symbolic and "informal" diagrammatic reasoning can complement one another in order for such a system to solve problems in a more understandable, intuitive and efficient way. Good sources of examples that indicate how extensive this research area has become are Narayanan 1992, Chandrasekaran et al. 1995, Anderson et al. 2000 and Anderson et al. 2001.

One of our aims is to explore the role of diagrams in mathematical proofs. We want to prove theorems of mathematics by manipulations of a diagram which capture the inference steps of a mathematical proof. The system we aim to implement proves theorems diagrammatically, and shows formally that the proofs are correct. The hope is that the truth and understanding of the proof remain transparent to the user of such a system through various combinations of diagram manipulations in the process of constructing proofs of theorems. We want to show that diagrams can be rigorous enough to be used for formal proofs. Moreover, we hope to capture in our system some of the intuitiveness and understanding of a proof which uses diagrams.[13]

The examples of theorems proved by manipulations of diagrams that we present in this chapter are our starting point for the investigation of the use of diagrams in formal proofs. In §3.5 we choose the domain of problems that we concentrate on in our pursuit to automate of diagrammatic reasoning in mathematics. As well as exploring the formality of diagrams in proofs, we also want to challenge Penrose's claim that diagrammatic reasoning cannot be automated for emulation on machines. We present here some of the type of diagrammatic reasoning that Pen-

[13]While the construction of diagrammatic proofs is intuitive and models how some humans reason with diagrams, the verification that such proofs are correct and formal (see Chapter 8) may be less so, since we have to use some symbolic logical tools to show the correctness of diagrammatic proofs. However, the verification is not something we use to construct diagrammatic proofs, but something that justifies the method we do use for constructing such proofs.

rose described. As already mentioned, Penrose presented his view in the lecture at International Centre for Mathematical Sciences in Edinburgh in November 1995. In Penrose 1994a he discusses in greater detail his disbelief in the possibility that computers will emulate reasoning with diagrams in any meaningful way, because mathematical visualization lies beyond any kind of purely computational activity. Our work could be seen as an attempt to disprove Penrose. However, this is not the only motivation for this book. Our work also explores the possibility of emulating diagrammatic reasoning on machines, and in fact, automates a small subset of it.

3.2 'Diagrammatic' Theorems

In order to clarify what we mean by diagrammatic proofs we first give some examples. We analyze these and devise a taxonomy, which helps us characterize the domain of problems under consideration.

Most of the examples presented here are taken from Nelsen 1993 and Nelsen 2001. These are an excellent source of numerous examples of proofs without words, or in Gardner's terminology, the "look-see" proofs. Nelsen's books are a collection of proofs without words from Ancient China, classical Greece, and twelfth-century India, but most of them are more recent. Frequently they appeared in the *Mathematical Association of America* journals. Other examples can be found in Dudeney 1958, Gamow 1988, Lakatos 1976, Gardner 1981, 1986 and Penrose 1994a.

For the examples that are presented here, we give the symbolic statement of the theorem first. Then, we show the diagrammatic representation of the theorem together with the geometric operations of the diagrammatic proof. Finally, we describe how the diagrammatic proof is carried out. Note that the descriptions of diagrammatic proofs are not necessary, because the proofs can be understood just by analyzing the diagrams. Therefore, the reader is invited to look at the picture representing a mathematical statement and try to see why it is true without reading the explanation under the picture.

3.2.1 Commutativity of Multiplication

The *commutativity of multiplication* theorem states that the order in which you multiply two numbers does not matter (see Figure 7).

The diagram that we present in Figure 7 is for any real number a and b. The diagrammatic proof would be the same if the theorem was for natural number arithmetic, and the diagrams would be represented as a collection of dots. The diagrammatic proof goes as follows: take a rectangle of any length a and height b. This represents a multiplication $a \times b$. Rotate this rectangle by 90 degrees. This results in a rectangle of

$$a \times b = b \times a$$

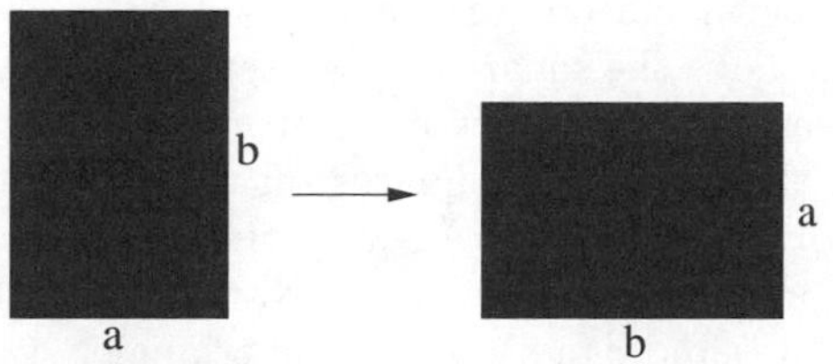

FIGURE 7 Commutativity of multiplication.

length b and height a, which represents a multiplication $b \times a$. The area of the rectangle is clearly preserved, hence $a \times b = b \times a$. Note that this is true for any values a and b.

3.2.2 Pythagoras' Theorem

Pythagoras' Theorem states that the square of the hypotenuse of a right-angle triangle equals the sum of the squares of its other two sides. Figure 8 shows one of the many different diagrammatic proofs of this theorem, taken from Nelsen 1993, page 3 (we give another example of a diagrammatic proof of Pythagoras' Theorem in Appendix A).

$$a^2 + b^2 = c^2$$

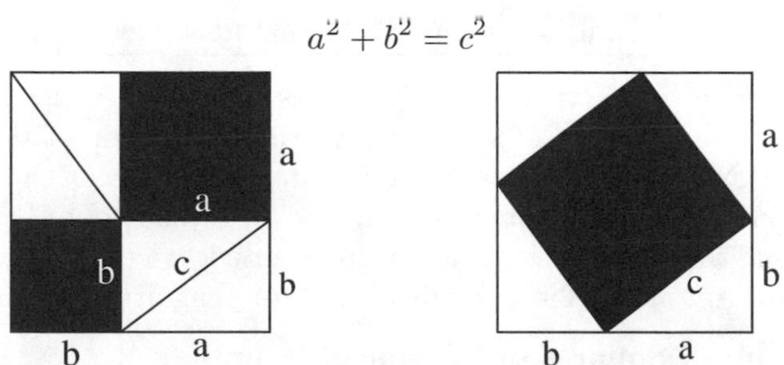

FIGURE 8 Pythagoras' theorem.

The proof consists of first taking any right-angle triangle. Along the hypotenuse c, join this triangle to another identical right-angle triangle, to make a rectangle. Join to this a square on the longer side a of one triangle, and a square on the shorter side b of the other triangle. Join to both squares, along their adjacent sides another two identical original right-angle triangles, but rotated 90 degrees. This completes the big square.

Now re-arrange the triangles into the big square so that each side of the square is formed from one side of one and the other side of another triangle. Thus, the magnitude of the big square is preserved and the square in the middle is the square on the hypotenuse. Clearly, when we subtract the areas of the four triangles from the original big and the new big square, the sum of the squares on the sides of the right-angle triangle (in the original big square) equals the square on the hypotenuse of this triangle (in the new big square). Notice again, that this is true for any values of a, b and c in a right-angle triangle.

3.2.3 Sum of Odd Naturals

This example is also taken from Nelsen 1993, page 71. The theorem about the *sum of odd naturals* is stated in Figure 9.

$$n^2 = 1 + 3 + \cdots + (2n - 1)$$

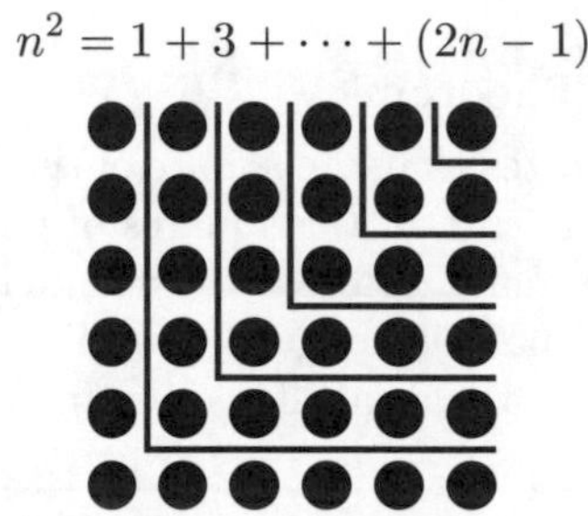

FIGURE 9 Sum of odd naturals.

If we take a square we can cut it into as many ells (which are made up of two adjacent sides of a square) as the magnitude of the side of the square. Note the use of parameter n in the number of applications of geometric operations. Note also that one ell is made out of two sides, i.e., $2n$, but the shared vertex has been counted twice. Therefore, each ell has a magnitude of $(2n - 1)$, where n is the magnitude of the square.

3.2.4 Sum of Squares of Fibonacci Numbers

The theorem about the *sum of squares of Fibonacci numbers* states that the sum of n squares of Fibonacci numbers equals the product of the n-th and $(n + 1)$-th Fibonacci number (see Figure 10). The example is taken from Nelsen 1993, page 83. The formal recursive definition of Fibonacci numbers is given as:

$$\begin{aligned}
Fib(0) &= 0 \\
Fib(1) &= 1 \\
Fib(2) &= 1 \\
Fib(n+2) &= Fib(n+1) + Fib(n)
\end{aligned}$$

$$Fib(n+1) \times Fib(n) = Fib(1)^2 + Fib(2)^2 + \cdots + Fib(n)^2$$

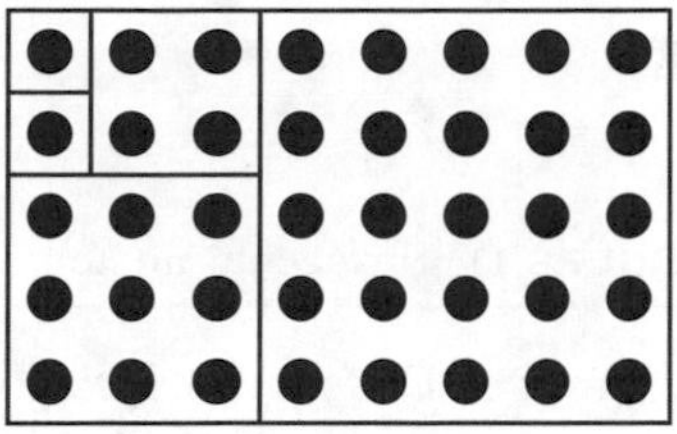

FIGURE 10 Sum of squares of Fibonacci numbers.

The diagrammatic proof in Figure 10 consists of taking a rectangle of length $Fib(n+1)$ and height $Fib(n)$ for some particular n. Decompose this rectangle by splitting from it a square of magnitude $Fib(n)$, which is the magnitude of the smaller side of the rectangle. Continue decomposing the remaining rectangle in a similar fashion until it is exhausted, i.e., for all n. Note the use of parameter n. Note also that the sides of the created squares represent consecutive Fibonacci numbers. Clearly, the longer side of every new rectangle is equal to the sum of the sides of two consecutive squares, which is precisely how Fibonacci numbers are defined.

This proof can be carried out inversely. We first take a square of unit magnitude (i.e., $Fib(1)^2$) and join it on one of its sides with another square of unit magnitude (i.e., $Fib(2)^2$). Therefore, a rectangle has been created. Take this rectangle and join to it a square of the magnitude of its longer side. A new rectangle will be created. Repeat this procedure for all n.

3.2.5 Sum of Hexagonal Numbers

The theorem about the *sum of hexagonal numbers* states that the sum of n hexagonal numbers equals n cubed:

$$n^3 = Hex(1) + Hex(2) + \cdots + Hex(n)$$

Hexagonal numbers can be formally defined by the recursive definition:

$$\begin{aligned} Hex(0) &= 0 \\ Hex(1) &= 1 \\ Hex(n+1) &= Hex(n) + 6 \times n \end{aligned}$$

The definition of hexagonal numbers could be presented in a series of hexagons where the hexagonal number is the number of dots in a hexagon (see Figure 11).

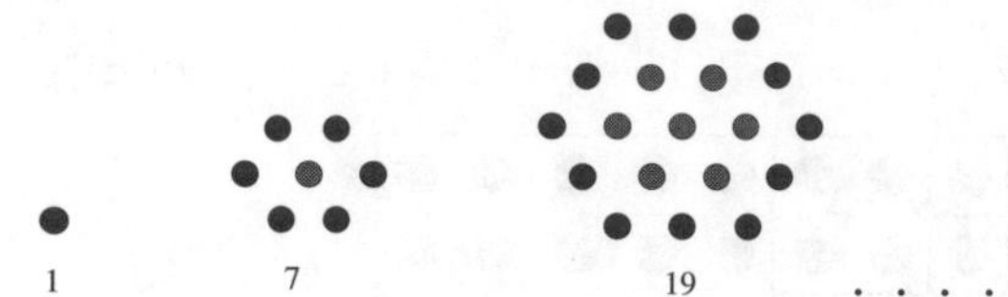

FIGURE 11 Hexagonal numbers.

The diagrammatic proof of the sum of hexagonal numbers consists of breaking a cube into a series of half-shells. A half-shell consists of three adjacent faces of a cube (see Figure 12). The example is taken from Nelsen 1993, page 109, and Penrose 1994a, pages 118-121.

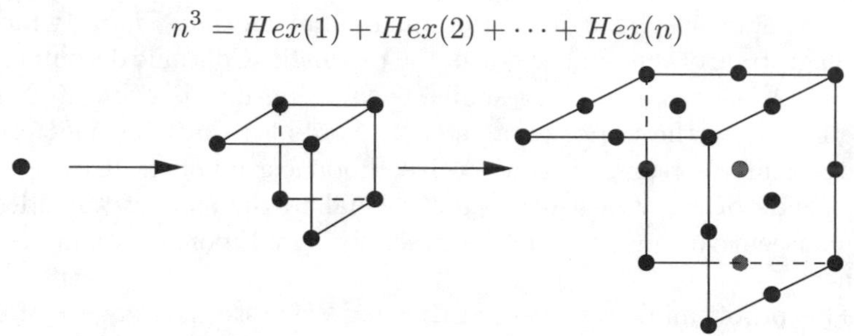

FIGURE 12 Sum of hexagonal numbers – half-shells.

If each half-shell is projected onto a plane, that is, if we look at the top-right-back corner of each half-shell down the main diagonal of the cube from far enough, then a hexagon can be seen. So the cube is then presented as the sum of all half-shells, i.e., hexagons (see Figure 13).

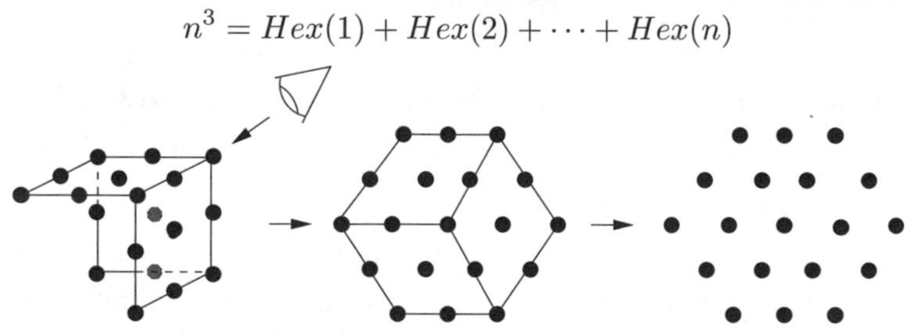

FIGURE 13 Sum of hexagonal numbers – project to 2D.

3.2.6 Triangular Equality for Even Squares

The following is a theorem about the *equality of triangular numbers for even squares.* A triangular number is defined to be $Tri(n) \equiv 1+2+3+\cdots+n = \frac{n(n+1)}{2}$. The example is taken from Nelsen 1993, page 101. The theorem states the following:

$$(2n)^2 = 8Tri(n-1) + 4n$$

Note that were we not to use the definition of triangular numbers, the theorem could be stated as $(2n)^2 = 8(1+2+3+\cdots+(n-1)) + 4n$. The diagrammatic proof is given in Figure 14.

$$(2n)^2 = 8Tri(n-1) + 4n$$

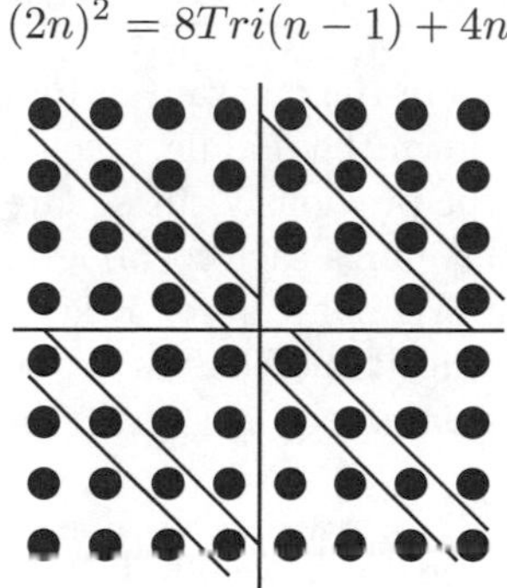

FIGURE 14 Triangular equality for even squares.

The proof consists of taking a square of magnitude $2n$ for a particular value of n. We then split it into four squares. Note that each of these four squares is of magnitude n. Split each of these four squares diagonally. For each square, two triangles are formed, one of magnitude n and one of magnitude $n-1$. For the four triangles of magnitude n, split from them one side. Note that the triangles become of magnitude $n-1$ and the sides are of magnitude n. Thus we have eight triangles of magnitude $n-1$, hence $8Tri(n-1)$ and four sides of magnitude n, hence $4n$. Notice that this is true for any natural number n.

3.2.7 Geometric Sum

This example is also taken from Nelsen 1993, page 118. The theorem is about a *geometric sum* of $\frac{1}{2^n}$ as n tends to infinity and is stated in Figure 15.

$$1 = \frac{1}{2} + \frac{1}{4} + \frac{1}{8} + \cdots$$

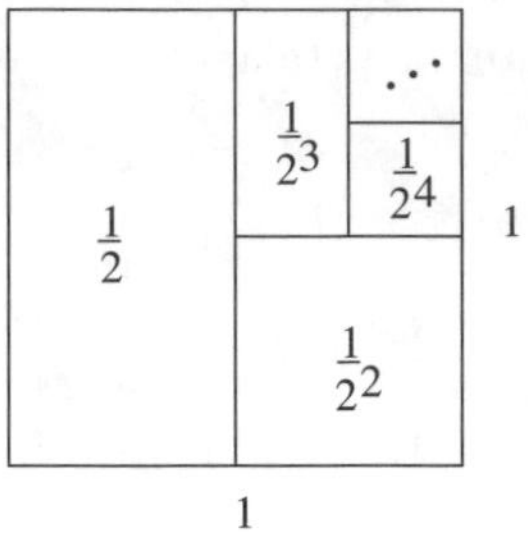

$\frac{1}{2}$ $\frac{1}{2^3}$ $\frac{1}{2^4}$ $\frac{1}{2^2}$ 1 1

FIGURE 15 Geometric sum.

Note the use of ellipsis in the diagram in Figure 15. Take a square of unit magnitude. Cut it down the middle. Each half has now the area of $\frac{1}{2}$. Now, cut one half of the previously cut square into halves again. This will create two identical squares each forming an area of $\frac{1}{2\times 2} = \frac{1}{2^2} = \frac{1}{4}$ making up a half of the original square. Take the top one of these two squares and repeat cutting it into halves as before indefinitely, which completes the original square.

3.2.8 Geometric Series

This example is also taken from Nelsen 1993, page 121. The theorem about a *geometric series* of $\frac{1}{(2^n)^2}$ (or equivalently $\frac{1}{4^n}$) as n tends to infinity is stated in Figure 16.

$$\frac{1}{3} = \frac{1}{4} + \frac{1}{16} + \frac{1}{64} + \cdots$$

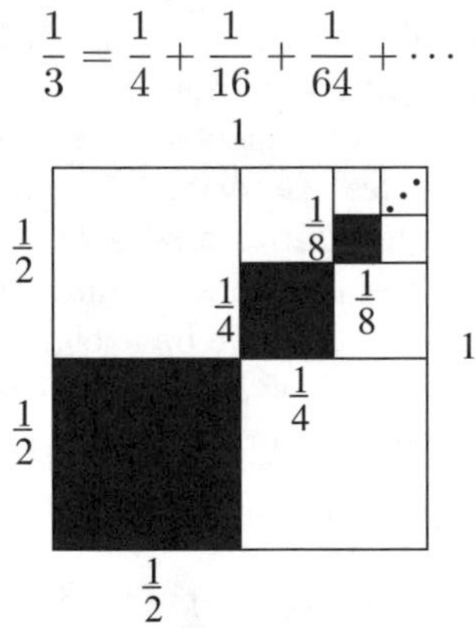

FIGURE 16 Geometric series.

Note the use of ellipsis in the diagram in Figure 16. Take a square of unit magnitude. Cut it into four squares. Note that each of the four squares is of magnitude $\frac{1}{2}$, thus the area of one of the four squares is $\frac{1}{4}$. Take one of the four squares and repeat the procedure. Note that this leaves three squares on which the procedure is not repeated (in the diagram above they form a kind of ell shape). The area of each newly created square is now $\frac{1}{4} \times \frac{1}{4} = \frac{1}{16}$. Continue to carry out the same procedure indefinitely. Note that the black squares are a third of the three squares on which the procedure is not repeated. As the number of such three-square structures tends to infinity, they comprise the entire original square of unit magnitude. Thus, the sum of all black squares is a third of the unit square.

3.3 Classification

We aim to choose from the examples represented in this chapter (and many more, some of which can be found in Appendix A) the class of theorems for which the construction of diagrammatic proofs will be automated. The classification of examples requires depicting certain properties of the examples and deciding the importance of each property. The examples are then evaluated and compared according to these properties, and finally classified into categories of common features.

The features which we consider interesting are:

- the domain of problems (e.g., continuous or discrete space, i.e., theorems about real or natural numbers),
- concreteness versus generality of diagrams,
- the need for abstraction devices in the representation of a diagram (this is directly related to the previous feature),
- the need for mathematical induction to prove the general case of the theorem,
- the number of proof steps.

These properties are by no means exhaustive, but they are the ones which help us to categorize the examples presented here.

We distinguish between a continuous space and discrete space of problems. Continuous space allows reasoning about real numbers, whereas discrete space only allows reasoning about natural numbers.

By concreteness of diagrams we mean the property that when a diagram is drawn it assumes a concrete magnitude, i.e., it represents particular values. By generality of a diagram we differentiate between diagrams that are the most general, i.e., they represent the whole class of diagrams, and diagrams that represent only an instance of a particular class of diagrams.

Depending on the above property, i.e., if a diagrams is concrete, then we need to use *abstraction devices* such as ellipsis to represent a general diagram. In a continuous space abstraction devices can be in the form of labelling a diagram. For instance, if a right-angle triangle is drawn in a continuous space, then it inherently assumes a concrete magnitude. Each side of the triangle could be labelled with some variable which indicates that the variable can assume any real value. Thus, a particular right-angle triangle is a representative of any right-angle triangle. The concreteness of diagrams is more problematic in a discrete space where diagrams are represented with points (or dots or counters etc.) on a grid. A diagram in a discrete space is an instance of the class that it is a part of. The generality of a diagram in a discrete space can be represented with the use of abstraction devices such as ellipsis. We discuss the difficulty of using abstraction devices in discrete space in §3.4.

Some theorems need *mathematical induction* to prove them. These are usually universally quantified over some parameter. We distinguish between theorems that have and those that do not have a notion of a universally quantified variable. Moreover, we are interested in theorems that are universally quantified over *one* parameter.

Some proofs of theorems consist of a number of proof steps dependent upon the instance (i.e., the value of the parameter) for which they are given. Such proofs are called *schematic proofs*. We distinguish between proofs that are schematic and those that are not. We discuss and formally define schematic proofs in Chapter 4.

The properties just discussed give a basis for the analysis of examples given in the previous section. The analysis will enable us to devise a taxonomy of theorems that admit diagrammatic proofs.

3.3.1 Analysis

Theorems about the *commutativity of multiplication* , *Pythagoras' theorem, geometric sum* and *geometric series* are theorems of continuous space. Diagrams in the proofs are represented using lines and sometimes also the coloring of areas. The main invariant of diagrams which is appealed to in order to convey proofs is the manipulation of diagram areas. For instance, the proof of *commutativity of multiplication* appeals to the fact that if we rotate the diagram by 90 degrees, its area remains the same.

Proofs of *commutativity of multiplication* and *Pythagoras' theorem* use diagrams which are general, i.e., they are representative of the entire class which represents the theorem. In other words, there is only one diagram for all instances of a theorem. There is no need to use ellipsis to represent the general case of a theorem. For example, a rectangle in the proof of *commutativity of multiplication* is representative of a rectangle of any magnitude, i.e., a and b stand for any real values. Hence, the labelling of a specific length with a variable can be considered an abstraction device, which is simpler to manipulate than the ellipsis. The same is true for a right-angle triangle in the *Pythagoras' theorem.* There is no mathematical induction needed to prove the general case of the theorem. Generalisation is required in the end to show that the theorem holds for all values of universally quantified variables. For both of these theorems, i.e., *commutativity of multiplication* and *Pythagoras' theorem*, the number of proof steps does not depend on any parameter, it is fixed.

On the other hand, proofs of *geometric sum* and *geometric series* do need abstraction devices to represent the theorem. In fact, there is no notion of instances of the theorem, because there is no universally quantified variable in the theorem. Therefore, there is only one case of a diagram, the one which represents the theorem. This case requires mathematical induction to prove the theorem. We say that theorems like these are inherently inductive. The number of diagrammatic operations is infinite.[14] Note however, that a universally quantified variable could be introduced, which would allow constructions of a diagrammatic proof. For example, the theorem about the *geometric sum* would instead of $1 = \frac{1}{2}+\frac{1}{4}+\frac{1}{8}+\cdots$ be stated for all $n > 0$ as $1 = (\frac{1}{2^1}+\frac{1}{2^2}+\frac{1}{2^3}+\cdots+\frac{1}{2^n})+\frac{1}{2^n}$. Figure 17 shows a diagrammatic proof for a particular instance, $n = 4$.

$$1 = (\frac{1}{2^1}+\frac{1}{2^2}+\frac{1}{2^3}+\cdots+\frac{1}{2^n})+\frac{1}{2^n}$$

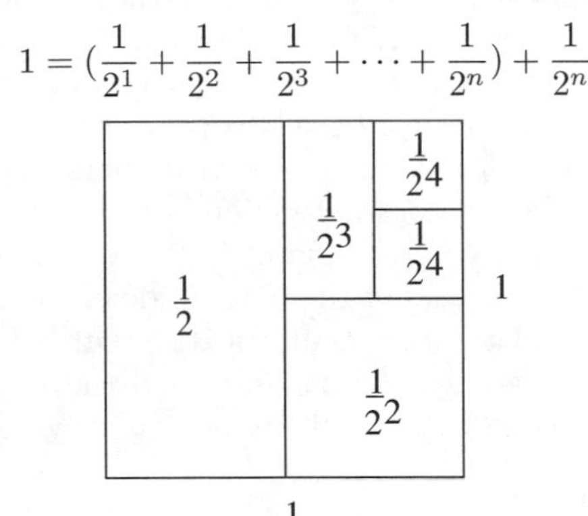

FIGURE 17 Universally quantified geometric sum.

Note that there is no longer any need for abstraction devices in the representation of an instance of the theorem and its proof. The same

[14]On the other hand, such theorems do admit finite symbolic (as opposed to diagrammatic) proofs.

could be done for the theorem about the *geometric series* — the new version of the theorem for all n would be: $\frac{1}{3} = (\frac{1}{(2^1)^2} + \frac{1}{(2^2)^2} + \frac{1}{(2^3)^2} + \cdots + \frac{1}{(2^n)^2}) + \frac{1}{(2^n)^2}$. However, such an introduction of a universally quantified variable transforms the theorem into a different one from the original.

Theorems about *triangular equality for even squares* , *sum of odd naturals, sum of squares of Fibonacci numbers* and *sum of hexagonal numbers* are theorems of a discrete space, in fact, the natural numbers. Diagrams that represent theorems and their proofs are drawn using dots, where a dot represents the natural number 1. An empty diagram, i.e., no diagram, is the number 0. The main feature of diagrams which is used to convey proofs is the manipulation of dots and its effect on the numbers that particular collections of dots represent.

The diagrams representing the proofs of these four theorems of discrete space are instances of a corresponding theorem. The universally quantified variable has been instantiated to a value and the diagram is drawn for this particular value. The diagram which is a representative of a particular instance of a theorem does not need abstraction devices. However, were we to represent a general case of a theorem, then an abstraction device would be needed in the diagram (an example of this is given in §3.4 in Figure 18).

For theorems about *sum of odd naturals* , *sum of squares of Fibonacci numbers* and *sum of hexagonal numbers*, the number of proof steps is dependent on the particular value for which the diagram is drawn, i.e., the value of a parameter for which the theorem is instantiated. The theorems are inductive, i.e., inductive consequences of axioms. Typically, mathematical induction may be used to prove the general case of the theorem. For the theorem about a *triangular equality for even squares* the number of proof steps does not depend on the value of the parameter for which the instance of the proof is given – the number of proof steps is constant. Abstraction of the magnitude of the discrete diagram is required in the end to show that the theorem holds for all values of the parameter. In a way, this is similar to the theorem about the *commutativity of multiplication* and *Pythagoras' theorem.*

3.3.2 Taxonomy

From the analysis of the examples presented in §3.2, Appendix A and many others, three categories of theorems and their proofs can be distinguished:

Category 1: Non-inductive theorems. Usually, there is only one representative diagram for all instances of the theorem. There is no need for mathematical induction to prove the general case. Proofs are not schematic, because the number of proof steps does not depend on the instance of the theorem. Simple geometric manipulations of the diagram prove the individual case. Generalisation is required to show that this proof holds for all a, b. Theorems are of continuous space. Example theorems: *commutativity of multiplication, Pythagoras' theorem.*

Category 2: Inductive theorems with a parameter. Each diagram represents a particular instance of a theorem. Theorems are inductive, so typically mathematical induction can be used to prove the general theorem for n (a concrete diagram cannot be drawn for this case). An alternative method, namely the constructive ω-rule (explained in the next chapter), can sometimes be used to capture the generality of the proof. Proofs are schematic, i.e., the number of proof steps depends on the instance of a theorem. Theorems are of discrete space. Example theorems: *triangular equality for even squares, sum of odd naturals, sum of squares of Fibonacci numbers, sum of hexagonal numbers.*

Category 3: Theorems whose proofs are inherently inductive: for each individual concrete case of the diagram they need an inductive step to prove the theorem. Every particular instance of a theorem, when represented as a diagram requires the use of abstraction devices to represent infinity. Theorems are of continuous space. Example theorems: *geometric sum, geometric series.*

Note that these categories are not exhaustive. We choose these, because they conveniently enable us to define our problem domain.

3.4 Abstraction Devices for Representation of Diagrams

Abstraction devices, such as ellipsis, labelling with variables and special signs such as $\sum$ for summing numbers, are conventions and notations which are used to represent generality or abstraction of a structure. They can be used in symbolic reasoning (e.g., $n^2 = 1+3+\cdots+(2n-1)$) or in diagrammatic reasoning. For example, were we to give the most general representation of a theorem about the *sum of odd naturals* and its proof, we would need to use ellipsis to represent a general diagram and a general number of applications of geometric operations on a diagram. Figure 18 shows the representation of a general square and the operations forming a proof of the theorem about the *sum of odd naturals.*

The fact that diagrams are concrete in nature is an inherent problem in drawing diagrams, which can be avoided by using ellipsis and other abstraction devices, as a convention to represent generality. However, the

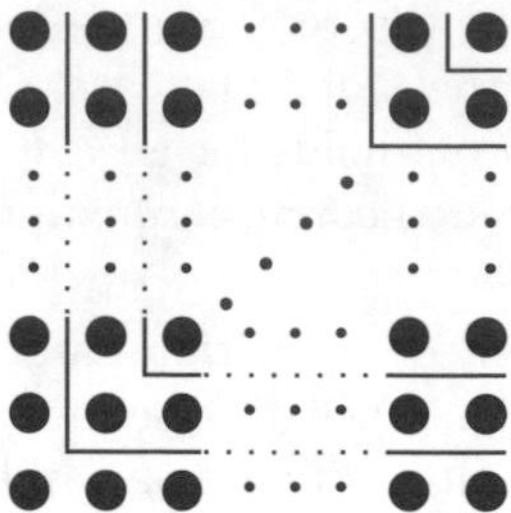

FIGURE 18 Abstract representations of a square in the proof of the *sum of odd naturals.*

problem in using such general diagrams is to keep track of what has been elided in the representation of a general diagram. This makes it difficult to count how many more occurrences of geometric operations still need to be applied. Note that general versions of geometric operations also need abstraction devices to represent their generality.

In symbolic representation there are formalizations of abstraction devices which are tractable throughout the manipulations in the proofs. For instance, the symbolic representation of the theorem about the *sum of odd naturals* is often expressed as: $n^2 = \sum_{i=1}^{n}(2i - 1)$ where the definition of $\sum$ is given recursively. Variables and other constructors such as $\sum$ are the abstraction devices.

In diagrammatic representations, particularly in a discrete space, the formalization of abstraction devices seems to be more difficult. This is one of the topics that needs addressing in the future. The problem lies in the manipulation of such abstraction devices. It is difficult to see how to automatically keep track of the consequences of operations being applied to the elided parts of the general diagrams. The inherent problem with ellipsis is its ambiguity. The pattern on either end or in the middle of the ellipsis needs to be induced by the system. For instance, it is ambiguous whether the general square given in Figure 18 is in fact a square or a rectangle. Some ambiguities can be removed by adding additional clues such as giving another layer of a diagram and having each corner of a general square be instantiated to a square. However, most of the ambiguities remain: is the square in Figure 18 of even magnitude or is it of odd magnitude? The problem becomes more acute when dealing with more complex structures. To recognize the pattern that the ellipsis represents, a computer system needs to carry out some sort of pattern recognition technique which deduces the most likely pattern and stores it in an exact internal representation. This guessed pattern might still be wrong.

There is a possibility to resort to a different abstract notation of diagrams which uses the exact internal representation rather than ambiguous ellipsis. The exact notation which would normally have to be deduced by the pattern recognition mechanism could be used in reasoning with an *internal* representation of abstraction devices. *Externally*, to the user of the system, this exact formalization could be made visual through a sort of pretty-printing technique. For instance, a general square of magnitude n which is given in Figure 18 and uses ambiguous ellipsis could internally be stored using an exact representation $square(n)$. All the internal reasoning can be carried out using this exact representation, yet the pretty-printing function would display to the user a square with ellipsis. The computational difficulty of extracting a pattern from an abstract notation has in this way been passed to the pretty-printing function.

Using such exact representation to store internally general diagrams and externally portray them using abstraction devices is open to many objections. The question can arise of how diagrammatic or non-diagrammatic this exact representation is. Are we not in essence carrying out symbolic reasoning which is the same as using n^2 instead of $square(n)$?[15] Where is the border which divides symbolic and diagrammatic reasoning, especially when automated on machines?

Symbolic v. Diagrammatic Representation

Trying to establish what is diagrammatic (or graphical) and what is symbolic in reasoning with a computer has been a topic of discussion amongst scientists in the fields like cognitive science, cognitive psychology, philosophy, computer science and artificial intelligence for a while (e.g., see Narayanan 1992, Olivier 1996, Blackwell 1997, Anderson 1997, Anderson et al. 2000, Anderson et al. 2001). There are views that *reasoning* entails linguistic processes. Therefore, even if we reason using diagrams, the diagram representation necessarily makes contact with language. Furthermore, a whole new area of programming using visual languages has been established (see Burnett and Baker 1994). Yet, we have not come closer to defining any precise distinction between the two. It seems that scientists adopt a distinction which is suitable within the

[15] Of course, any reasoning by a computer system could be considered as symbolic, because everything needs to be represented at the lowest level in terms of symbols on a computer. However, we do not support this view, we consider reasoning on different levels of abstraction. Therefore, we distinguish between reasoning with, for example, a diagram that is a square and a symbolic expression n^2.

scope of their research. We adopt here an informal notion that $square(n)$ and n^2 are symbolic representations, because they have no properties that are analogous to our visual comprehension of a square. On the other hand, the representation of a square given in Figure 18 is considered to be diagrammatic. Furthermore, if the constructed proof consists of no symbolic inference rules, then we consider it a diagrammatic proof, even though it is a diagrammatic proof of some symbolical conjecture.

3.5 Problem Domain

In §3.2 we introduced the notion of diagrammatic theorems through a number of examples. We discussed in §3.3 their common features which enabled us to categorize them. The categorization is by no mean exhaustive, but it helps us choose our problem domain.

First, we choose mathematics as our domain for theorems since it allows us to make formal statements about reasoning, proof search, induction, generalizations, abstractions and such issues. All of these are important when automating a system that carries out the type of diagrammatic reasoning represented by the examples in §3.2.

Second, we narrow down the domain to a subset of theorems that can be represented as diagrams without the need for abstraction devices. Conducting proofs and using abstraction devices in diagrams is problematic, as explained in §3.4, since it is difficult to keep track of these abstractions while manipulating the diagram during the proof procedure. This excludes the whole of theorems of Category 3 in §3.3.2. We also reject representing general diagrams of Category 2 (as the one in Figure 18), but only concrete (i.e., instantiated) versions of them (as the one in §3.2.3). Therefore, theorems of Category 2 will be instantiated to particular values, and the reasoning will be carried out on these instances. The generality of the proof will be captured in an alternative way.[16]

Third, we consider inductive theorems. These may typically be proved in the general case with mathematical induction (i.e., Category 2 given in §3.3.2). Such theorems are universally quantified over some parameters – we consider only the ones that are universally quantified over one parameter. This includes theorems of Category 1 and Category 2. Inductive proofs reason about general values of parameters which suggests that theorems need to be represented with general diagrams containing abstraction devices. For example, an $n \times n$ square in Figure 18 cannot be drawn without using ellipsis. Our challenge is to find a mechanism

[16]We use schematic proofs, which will be introduced in Chapter 4, to capture the generality of the proof.

for constructing a general proof that does not require using diagrams with ellipsis. The generality of the proof will be captured in a different way.

Fourth, to date we consider theorems of natural number arithmetic only. This area is rich, interesting and different to other research done in this field (see the survey of diagrammatic reasoning systems in Chapter 2), because arithmetic theorems are not as obviously amenable to diagrammatic representations as geometric theorems are. Diagrams that represent theorems of natural number arithmetic are represented using dots. The problem space is two or three dimensional and discrete. In this book we only present diagrammatic proofs of theorems that are expressed in two dimensions, but the approach generalizes to three dimensions as well, as will be shown later. Notice that the domain of theorems that we can prove diagrammatically is not limited to only theorems which are expressed as degree two or three polynomial equations, and which have an obvious two or three dimensional diagrammatic representation. We give here an example of a theorem which is stated using an equation of degree three polynomial, yet the diagrammatic proof uses diagrams of a two dimensional space only. The theorem is about the *sum of cubes* and is given with its diagrammatic proof for $n = 4$ (the example is taken from Nelsen 1993, page 85) in Figure 19.

$$(1 + 2 + 3 + \cdots + n)^2 = 1^3 + 2^3 + 3^3 + \cdots + n^3$$

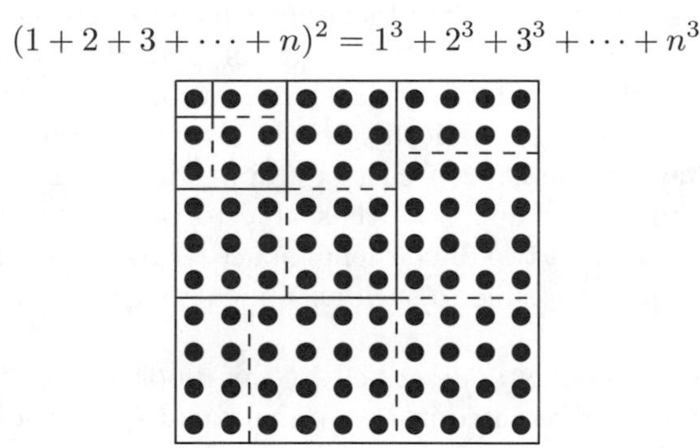

FIGURE 19 Sum of cubes.

The diagrammatic proof of this theorem in Figure 19 consists of taking a square of magnitude $(1+2+3+\cdots\ +n)^2$ for some particular n (in the case of $n = 4$ the square is of magnitude $1 + 2 + 3 + 4 = 10$) and splitting it into strips of ells each one unit thicker than the previous one. This explains why the original square is of magnitude

$(1+2+3+\cdots+n)^2$. For each ell of thickness t we split it now into as many squares of magnitude t as possible. For instance, an ell of thickness 3 can be split to three squares of magnitude 3. Thus $3\times 3^2 = 3^3$. Note that for each ell of even thickness k, only $k-1$ squares of magnitude k fit into an ell, plus two bits at the end of an ell which form half of the square of magnitude k. Hence, both of them together form another square of magnitude k. So, $2\times \frac{k^2}{2} = k^2$. Therefore, for each ell of even thickness we have $((k-1)\times k^2)+(2\times \frac{k^2}{2}) = (k-1)\times k^2+k^2 = k\times k^2 = k^3$. To be more accurate, the diagrammatic proof given here proves the following version of the theorem: $(1+2+3+\cdots+n)^2 = 1\times 1^2+2\times 2^2+3\times 3^2+\cdots+n\times n^2$, because we are not appealing to any three dimensional property of a cube. A three dimensional version of this diagrammatic proof is to think of dots as spheres, and take for each thick ell, all of the sectioned squares and join them one on top of another to form a cube.

The example just given shows that the degree of the polynomial in a theorem does not uniquely determine the dimension of the space for a diagram representing this theorem. Another example which demonstrates this is a theorem which uses polynomials of degree four. Some humans find it difficult to picture four dimensional space, yet this does not prevent us from proving such theorems. For example, if we have a term n^4 we can alway represent it as $n\times n^3$ in a three dimensional space for some concrete n.

There are some limitations to theorems which we can prove diagrammatically. If we can appeal to the feature of a diagram which conveys the truth of the theorem in two or three dimensional space, then we can prove a theorem diagrammatically. In the example above, we proved the theorem about the *sum of cubes* by appealing to the fact that n^3 is equivalent to $n\times n^2$ and were thus able to prove the theorem in a diagrammatic way in a two dimensional space. Only theorems for which these features are accessible to such an appeal can be proved diagrammatically.

The system DIAMOND described in this book proves theorems of Category 2 which represented in two-dimensional space and are universally quantified over one parameter. The number of proof steps may be dependent on this parameter, thus the proofs are called schematic.

3.6 Summary

Diagrams have been used to aid reasoning throughout the history of mathematics. However, at the turn of the twentieth century, with the invention of modern logic, diagrams seemed to have lost their rigor in formal proofs. Recently, much research has been done to re-establish the formal role of diagrams. The work presented in this book is part of this trend. In this chapter, we introduced examples of theorems that admit diagrammatic proofs which we call diagrammatic theorems. The analysis of the diagram features in these examples enabled us to introduce a taxonomy of diagrams which is used to help us select our problem domain.

Some abstraction devices are inherently ambiguous when used in diagrams. However, there is scope to avoid them by a different type of notation that is exact and requires a sort of pretty-printing function to represent the proof externally in a diagrammatic way. However, it is questionable whether such notation can still be called diagrammatic, or has it been reduced to a symbolic notation. The debate about what is diagrammatic and what is symbolic is a long-standing one. Since there is no generally accepted definition for them, we use our own informal characterization, which enables us to distinguish between diagrammatic and symbolic proofs.

Finally, we chose our problem domain to be theorems of natural number arithmetic which require mathematical induction to prove them in the general case. We choose these, because they enable us to exploit their concreteness property, and represent them as collections of dots. The theorems are universally quantified over one parameter. The number of proof steps in such theorems is dependent upon the parameter. A subset of these theorems have proofs which are called schematic proofs – we are interested in these and hence explain them next.

4

The Constructive ω-Rule and Schematic Proofs

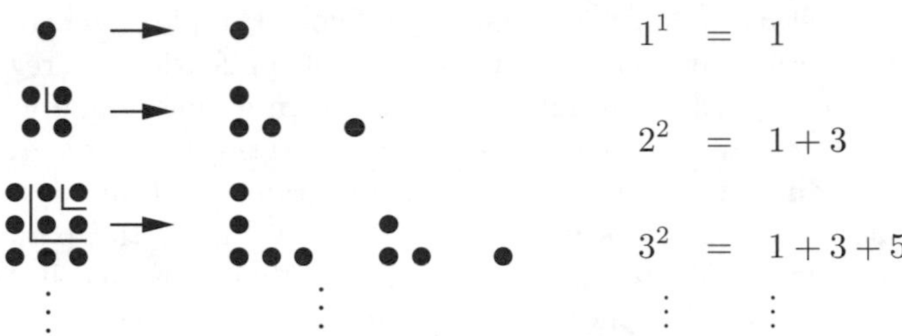

— MJ
adapted from NELSEN's *Proofs Without Words*

In the previous chapter we presented some of the examples of theorems which we prove diagrammatically. The topic of this chapter is to present a way of capturing diagrammatic proofs without the need to resort to general diagrams which use abstraction devices. The mathematical basis for capturing the generality of the proof is in the use of the constructive ω-rule in schematic proofs, which is explained in this chapter.

We start, in §4.1, by placing our choice of the technique for construction of diagrammatic proofs in the context of automated reasoning. We go on in §4.2 to explain the constructive ω-rule. One way of automating its use is by constructing schematic proofs, which we define next in §4.3. In §4.4 we explain how to construct schematic proofs and give an example of a schematic proof in arithmetic. There are a number of motivations for using schematic proofs and these are discussed next in §4.5. In §4.6 we challenge Penrose's argument that diagrammatic proofs cannot be automated. In §4.7, we propose how to use schematic proofs for the representation of diagrammatic proofs. Finally, in §4.8, we give structured informal schematic proofs for examples of Category 2 proofs given in Chapter 3.

4.1 Motivation

The theorems whose diagrammatic proofs we aim to formalize and automate, are typically proved by mathematical induction in a symbolic logical proof. Using mathematical induction in a diagrammatic proof requires the use of general diagrams with abstraction devices, which we are trying to avoid (see §3.4). An alternative way of capturing the generality of a diagrammatic proof of a theorem is to use schematic proofs.

We sketch here the basic idea behind schematic proofs, but define them fully in §4.3.[17] A schematic proof is a recursive program with some parameters. By instantiation of these parameters the program generates ground instances of a particular proof. For example, a schematic proof in arithmetic may consist of a number of applications of rewrite rules which are applied to an initial expression. In a diagrammatic proof the rewrite rules are replaced by geometric operations on a diagram. Thus, a diagrammatic schematic proof is a program which applies geometric operations to diagrams when given some value of the parameter. In this way, we eliminate the need for general diagrams, and instead use a general number of applications of geometric operations.

A schematic proof is constructed from a number of ground instances of a proof for a corresponding ground instance of a theorem. This process is referred to as inductive inference (Winston 1975) or abstraction,[18] because it induces general conclusions from particular instances, i.e., examples. The particular abstraction technique that we implement is discussed later in §7.5 – here we formalize the *algorithm* for constructing schematic proofs which are mathematically based on the constructive ω-rule.

4.2 Constructive ω-Rule

The constructive ω-rule is based on the ω-rule. The ω-rule is an infinitary logical rule which requires an infinite number of premises to be proved in order to conclude a universal statement. The definition of the ω-rule, the motivation for its use within logic, and an example of its use are given in Appendix B.

[17]The formalization of schematic proofs that represent diagrammatic proofs, and the implementation of the construction of schematic proofs is discussed in Chapter 7.

[18]For definitions of inductive inference and abstraction, see Glossary. Note the difference with respect to mathematical induction.

We restrict the ω-rule so that the infinitary proofs which are needed possess some important properties of finite proofs. One such restriction is the so called constructive ω-rule. This rule essentially requires that there is a recursive function which generates by instantiation all instances of a proof. A possible implementation of the recursive function, required by the constructive ω-rule, is by finding a general pattern of a proof from example-proofs for instances of a theorem, and capturing it in a recursive program. This can be done by an abstraction mechanism. An algorithm which can be used to recognize automatically the general pattern abstracts an initial set of rewrite rules describing an instance of a proof, and then updates this abstraction according to other instances of a proof, until the general proof representation satisfies all of the (large number of) cases considered. Any abstraction algorithm can be used to guess the ω-proof from individual proof instances – our choice is discussed in §7.5.

We can now formally define the constructive ω-rule, which we will use to prove diagrammatic theorems of Category 2[19] in a similar way that the rule is used to prove theorems of arithmetic. The constructive ω-rule and schematic proofs for theorems of arithmetic were investigated by Baker (1992).[20] Notice that ω is an ordered set of natural numbers, and s is a successor function (see Glossary).

Definition 1 (Constructive ω-Rule)
The constructive ω-rule allows inference of the sentence $\forall x.\ P(x)$ from an infinite sequence $P(n)$ for $n \in \omega$ of sentences

$$\frac{P(0), P(s(0)), P(s(s(0))), \ldots}{\forall n. P(n)}$$

such that each premise $P(n)$ is proved **uniformly** (from parameter n).

The uniformity criterion is taken to be the provision of a computable procedure describing the proof of $P(n)$. The requirement for a uniform procedure is equivalent to the notion that the proofs for all premises are captured in a recursive function. This means that a proof tree and a function describing the use of different rules in a proof need to be recursive. We call such uniform recursive functions *schematic proofs*.

4.3 Schematic Proof

Here is a formal definition of a schematic proof.

---

[19]The taxonomy of diagrammatic theorems was given in §3.3.

[20]More information on Baker's work can be found in Baker and Smaill 1995 and Baker 1993.

Definition 2 (Schematic Proof)
A schematic proof is a recursive function which outputs a proof of some proposition $P(n)$ given some n as input.

Suppose the recursive function, proof, is a schematic proof. The function proof takes one argument, namely a parameter n. By instantiation (i.e., by assigning a particular value to n and passing it as an argument to the function proof), and by application of this instantiated function to the theorem, $\mathsf{proof}(n)$ generates a proof for a particular premise $P(n)$. More informally, $\mathsf{proof}(n)$ describes the inference steps (i.e., rules) made in proofs for each $P(n)$. Now, $\mathsf{proof}(n)$ is schematic in n, because we may apply some rule R a function of n number of times. That is, the number of times that a rule R is applied in the proof might depend on the parameter n. This recursive definition of a proof is used as a basis for implementation of the schematic proofs (see §7.3).

4.3.1 Example of Schematic Proof in Arithmetic

To illustrate the use of the constructive ω-rule in schematic proofs, we give here an example of a schematic proof of a theorem of arithmetic. Consider a special version (with one universally quantified variable) of a theorem about the *associativity of addition*, stated as

$$(x+x)+x \quad = \quad x+(x+x) \tag{4.1}$$

Baker studied schematic proofs of such theorems in Baker et al. 1992. The recursive definition of plus is given as follows:

$$0+Y \quad = \quad Y \tag{4.2}$$

$$s(X)+Y \quad = \quad s(X+Y) \tag{4.3}$$

We also need a reflexive law $\forall A.\ A = A$.

The constructive ω-rule is used on x in the statement of the *associativity of addition*. We write any ground instance of x as $s^n(0)$. By $s^n(0)$ is meant the n-th numeral, i.e., the term formed by applying the successor function n times to 0. Next, the axioms are used as rewrite rules from left to right, and substitution is carried out in the ω-proof, under the appropriate instantiation of variables. We use the following two instances of the constructive ω-rule in our example:

$$\frac{\begin{array}{c}(s^2(0)+s^2(0))+s^2(0)=s^2(0)+(s^2(0)+s^2(0)),\\[1ex] (s^3(0)+s^3(0))+s^3(0)=s^3(0)+(s^3(0)+s^3(0))\end{array}}{\forall x. \quad (x+x)+x=x+(x+x)}$$

where the parameter n in the instance $s^n(0)$ of x is 2 and then 3. We construct a schematic proof in terms of the values of this parameter.

The aim is to reduce both sides of the equation to the same term. The schematic proof is the program proof:

$$\begin{aligned}\mathsf{proof}(n) = \quad & \text{Apply rule (4.3) } n \text{ times on both sides of equality,}\\ & \text{Apply rule (4.2) once on both sides of equality,}\\ & \text{Apply rule (4.3) } n \text{ times on left side of equality,}\\ & \text{Apply reflexive law}\end{aligned}$$

Running this program on the associativity theorem (4.1) proves it. For example (bold blocks represent program execution steps, i.e., applications of rewrite rules on the theorem), here is the proof for the instance $x = 2$, i.e. $s^2(0)$:

$$\begin{aligned}
(s^2(0) + s^2(0)) + s^2(0) &= s^2(0) + (s^2(0) + s^2(0))\\
&\Downarrow \ \textbf{Apply rule (4.3) on both sides}\\
(s^1(s^1(0) + s^2(0)) + s^2(0) &= s^1(s^1(0) + (s^2(0) + s^2(0)))\\
&\Downarrow \ \textbf{Apply rule (4.3) on both sides}\\
s^2(0 + s^2(0)) + s^2(0) &= s^2(0 + (s^2(0) + s^2(0)))\\
&\Downarrow \ \textbf{Apply rule (4.2) on both sides}\\
s^2(s^2(0)) + s^2(0) &= s^2(s^2(0) + s^2(0))\\
&\Downarrow \ \textbf{Apply rule (4.3) on left}\\
s^1(s^1(s^2(0)) + s^2(0)) &= s^2(s^2(0) + s^2(0))\\
&\Downarrow \ \textbf{Apply rule (4.3) on left}\\
s^2(s^2(0) + s^2(0)) &= s^2(s^2(0) + s^2(0))\\
&\Downarrow \ \textbf{Apply reflexive law}\\
&\mathit{true}
\end{aligned}$$

Here is the proof for the instance $x = 3$ coded as $s^3(0)$:

$$\begin{aligned}
(s^3(0) + s^3(0)) + s^3(0) &= s^3(0) + (s^3(0) + s^3(0))\\
&\Downarrow \ \textbf{Apply rule (4.3) on both sides}\\
(s^1(s^2(0) + s^3(0)) + s^3(0) &= s^1(s^2(0) + (s^3(0) + s^3(0)))\\
&\Downarrow \ \textbf{Apply rule (4.3) on both sides}\\
(s^2(s^1(0) + s^3(0)) + s^3(0) &= s^2(s^1(0) + (s^3(0) + s^3(0)))\\
&\Downarrow \ \textbf{Apply rule (4.3) on both sides}
\end{aligned}$$

$$
\begin{array}{rcl}
s^3(0 + s^3(0)) + s^3(0) & = & s^3(0 + (s^3(0) + s^3(0))) \\
& \Downarrow & \textbf{Apply rule (4.2) on both sides} \\
s^3(s^3(0)) + s^3(0) & = & s^3(s^3(0) + s^3(0)) \\
& \Downarrow & \textbf{Apply rule (4.3) on left} \\
s^1(s^2(s^3(0)) + s^3(0)) & = & s^3(s^3(0) + s^3(0)) \\
& \Downarrow & \textbf{Apply rule (4.3) on left} \\
s^2(s^1(s^3(0)) + s^3(0)) & = & s^3(s^3(0) + s^3(0)) \\
& \Downarrow & \textbf{Apply rule (4.3) on left} \\
s^3(s^3(0) + s^3(0)) & = & s^3(s^3(0) + s^3(0)) \\
& \Downarrow & \textbf{Apply reflexive law} \\
& \mathit{true} &
\end{array}
$$

In general, the proof for all instances of x, i.e., for all values of n in $s^n(0)$ can be represented as:

$$
\begin{array}{rcl}
(s^n(0) + s^n(0)) + s^n(0) & = & s^n(0) + (s^n(0) + s^n(0)) \\
& \Downarrow & \textbf{Apply rule (4.3) } n \textbf{ times on both sides} \\
& \vdots & \\
s^n(0 + s^n(0)) + s^n(0) & = & s^n(0 + (s^n(0) + s^n(0))) \\
& \Downarrow & \textbf{Apply rule (4.2) once on both sides} \\
s^n(s^n(0)) + s^n(0) & = & s^n(s^n(0) + s^n(0)) \\
& \Downarrow & \textbf{Apply rule (4.3) } n \textbf{ times on left} \\
& \vdots & \\
s^n(s^n(0) + s^n(0)) & = & s^n(s^n(0) + s^n(0)) \\
& \Downarrow & \textbf{Apply Reflexive Law} \\
& \mathit{true} &
\end{array}
$$

Note that the number of proof steps depends on n, which is the instance $s^n(0)$ we are considering. We see that the proof is schematic in n — certain steps are carried out a number of times depending on n.

We now show how schematic proofs of universally quantified theorems can be found using several heuristics.

4.4 Finding a Schematic Proof

A schematic proof can be constructed by considering individual example-proofs for instances of a theorem, and then extracting a general pattern from these instances. This general pattern can be captured in a recursive function, proof. The idea is that in order to construct a general structure common to all instances of a proof, the particular example-proofs of a theorem which are considered need to be general representatives of all instances, and not just a subset of them. These general representative instance are normally taken to be for some intermediate values, e.g., 5 and 6, or 7 and 9, rather than the initial values, e.g., 0 and 1, since the proofs for initial values of a parameter n are almost always special cases. Therefore, we use such intermediate values, e.g., $P(7)$ and $P(9)$ and correspondingly $\mathsf{proof}(7)$ and $\mathsf{proof}(9)$, to extract the pattern, which hopefully is general. That is, we hope that this pattern is representative of all instances of a proof. A structure which is common to the considered examples is constructed by an abstraction mechanism. The result is the construction of a general schematic proof. If the instances for the intermediate values that were considered are not typical for all instances, so that the abstraction mechanism was carried out on incomplete information, then the constructed recursive function proof could be wrong. Therefore, the function proof needs to be verified as correct. This involves reasoning about the proof (using meta-level reasoning), and showing that when proof is input any n and applied to a theorem, it indeed generates a correct proof of each $P(n)$.

The following procedure summarizes the essence of using the constructive ω-rule in schematic proofs:

1. Prove a few particular cases (e.g., $P(7)$, $P(9)$, ... and thereby discover $\mathsf{proof}(7)$, $\mathsf{proof}(9)$, ...).
2. Abstract $\mathsf{proof}(n)$ from these proofs (e.g., find a pattern from $\mathsf{proof}(7)$, $\mathsf{proof}(9)$, ...).
3. Verify that $\mathsf{proof}(n)$ proves $P(n)$ (e.g., by meta-induction on n).

The general pattern is abstracted from the individual proof instances by a learning process described in more detail in §7.5. We now explain the notion of meta-induction.

4.4.1 Meta-Induction for Verification of Schematic Proofs

By meta-mathematical induction we mean that we introduce a theory META such that for all n:

$$\vdash_{\text{META}} \mathsf{proof}(n) : P(n)$$

where ":" stands for "is a proof of". The meta-inductive rule is defined as follows:

$$\frac{\vdash_{\text{META}} \mathsf{proof}(0) : P(0) \qquad \mathsf{proof}(n) : P(n) \vdash_{\text{META}} \mathsf{proof}(s(n)) : P(s(n))}{\vdash_{\text{META}} \ \forall n \ \mathsf{proof}(n) : P(n)}$$

This essentially says that by using the rules on $P(s(n))$ we can reduce it to $P(n)$. By meta-induction we need to show in the meta-theory that given a proposition $P(n)$, $\mathsf{proof}(n)$ indeed proves it, i.e., it gives a correct proof with $P(n)$ as its conclusion, and axioms of some chosen logic as its premises. This ensures that the constructed general schematic proof is indeed a correct proof for all instances of a proposition. Meta-induction differs from standard mathematical induction in that it makes an assertion about proofs rather than object-level formulae.

In order to show in the meta-theory that $\mathsf{proof}(n)$ proves the proposition $P(n)$ we need to encode $P(n)$, so that the proposition is transformed from the object-level statement to the meta-level statement. This can be done in a theory of diagrams which will be discussed in Chapter 8.

4.5 Why Use Schematic Proofs?

We discuss here several informal motivations for using schematic proofs. That is, we make some conjectures about the psychological validity of schematic proofs. We have anecdotal evidence to support our intuitions, however, we have not conducted any systematic experiments. Such an investigation would shed some light on the nature of human mathematical thought.

4.5.1 Learning from examples

Schematic proofs and the constructive ω-rule explain why one or more examples can represent a general proof. Therefore, our first conjecture is that schematic proofs justify the use of examples for inducing formal proofs.

As described in §4.3, the constructive ω-rule enables us to capture infinitary concepts in a finite way, i.e., it enables us to use schematic proofs in order to prove universal statements. The constructive ω-rule gives us a mathematical basis which justifies how and why the examples or instances of problems can be used in order to conclude, in our case, a general proof of a universally quantified theorem. Schematic proofs have been used to reason with instances of universally quantified theorems of arithmetic (Baker and Smaill 1995), such as the theorem about *associativity of addition* given in §4.3.1.

We will propose in §4.7 to use schematic proofs for diagrammatic proofs of the kind we presented in Chapter 3, precisely because they allow us to use examples to construct general proofs. There is no longer a need for abstract diagrams which use ellipsis to represent generality. We can use concrete examples of diagrams and use schematic proofs to capture the generality by a variable number of applications of geometric operations on a diagram. The intricacies of how schematic proofs can be used for a formalization of diagrammatic proofs will be discussed in §4.7.

Besides the ability to construct general proofs from examples, it also appears that reasoning with examples seems easier for humans to understand than reasoning with abstract notions. That is, using instances of theorems to prove them may be easier and may convey better why a theorem holds. The usual way of proving the theorem about *associativity of addition* is to use mathematical induction and a generalization, which is difficult to find for a human and an artificial mathematician – a mechanized mathematical reasoning system.

4.5.2 Erroneous proofs

In an automated reasoning system, formality is important in order to ensure the correctness of the automatically derived conclusions. Hence, in a system that induces schematic arguments, their correctness has to be formally shown. This confirms that a schematic proof is indeed a correct formal proof of a theorem. If all proofs of theorems that people find followed rules of some formal logic, then there would be no explanation for how erroneous proofs could arise. The errors would always be detected as syntactical errors, provided that the rules used to prove the theorem are correct.

However, the history of mathematics has taught us that there are plenty of faulty proofs of theorems which were for a long time considered to be correct, but later it turned out that the "proofs" were not proofs at all, that is, they were incorrect. So what is going on, why do erroneous "proofs" persist? Our conjecture is that schematic proofs account for erroneous proofs.

One of the most famous examples of a faulty "proof" is the history of *Euler's theorem* (Lakatos 1976) and its erroneous proof.[21] *Euler's theorem* states that for any polyhedron $V - E + F = 2$ holds, where V is the number of vertices, E the number of edges, and F is the number of faces. Lakatos[22] initially gives a proof, historically due to Cauchy, of the theorem, which is a uniform program for proving instances of *Euler's theorem*. Thus, the program is a schematic proof. However parts

[21] Another example is a "proof" of the *4-color conjecture* (Kempe 1879) which had faulty proofs. The problem is in the formalization of a conjecture.

[22] The proof of *Euler's theorem* is also discussed in Gamow 1988, pages 47-48.

of the program are not explicitly stated, but seem very convincing when applied to simple polyhedra.[23]

Analyzing Cauchy's proof, Lakatos proceeds to present a number of counterexamples in which the program fails. It turns out that the initial theorem does not hold for *all* polyhedra. The reader is referred to Lakatos 1976 for a number of counterexamples for this theorem and its "proof". Cauchy used a schematic proof in order to convince us of his "proof" of *Euler's theorem*. However, he did not carry out the last step of the procedure for construction of schematic proofs (like the one given in §4.4), namely, he did not verify that the schematic proof is indeed correct. We argue that if he did use the complete procedure, then the fallacy of his schematic "proof" would be detected at the verification stage.

It seems plausible that humans use some sort of procedure for construction of schematic proofs to find general proofs of theorems. In particular, humans often use examples of proofs for certain instances and then abstract them into a general schematic proof. If not all the cases are covered by the examples, then the schematic proof might be incorrect, as in the case of the proof of *Euler's theorem* mentioned above. If a counterexample is encountered, then the program needs to be revised to exclude such cases. It seems that humans sometimes omit this verification step all together. Human machinery for constructing a general schematic argument is usually convincing enough to reassure them that their general schematic argument is correct, e.g., consider the "proof" of *Euler's theorem*. It is plausible that humans are happy with the intuitive understandings of definitions and steps in the proof – as long as they do not encounter a counterexample, their general schematic argument for the proof is acceptable to them. Lakatos refers to such mathematical proofs as "thought experiments". It is only recently, in the 20th century, that thought experiments were replaced by formal proofs.

So, our second conjecture is that human mathematicians often use a procedure similar to the construction of schematic proofs given in §4.4 in order to find proofs of theorems, but they often omit the verification step which ensures that the proof is correct. We propose further, that omitting the last step of such procedure accounts for numerous examples of faulty "proofs". For instance, if one has not considered all the representative examples, then the schematic proof may not prove all cases of the theorem. A counterexample could be found.

[23]See §A.5 for a full explanation of the proof procedure.

4.5.3 Intuitiveness of schematic proofs

Finally, we propose that schematic proofs seem to correspond better to human intuitive proofs. It appears easier to see why the theorem holds by looking at the instances of a theorem and its proof and then constructing a schematic proof, than considering a logical proof. One example was already discussed, namely Baker's proof of *associativity of addition*. Later we propose how to use schematic proof for constructing a diagrammatic proof of the *sum of odd naturals* (see §4.8.1). Both of these proofs seem to be easier to understand as schematic proofs than formal logical proofs. Another example is Penrose's *sum of hexagonal numbers* given in §3.2.5. We now present a further example which supports our conjecture, namely the *rotate-length* theorem.

The *rotate-length* theorem is about rotating a list its length number of times, and can be stated as:

$$rotate(length(l), l) = l$$

where $length(l)$ gives the length of a list l, and $rotate(x, l)$ takes the first x elements of a list l and puts them at the end of it (e.g., $rotate(3, [a, b, c, d, e]) = [d, e, a, b, c]$). Consider a schematic proof of this theorem. First we give an example-proof for the instance of a theorem with a list of any five elements $l = [a, b, c, d, e]$, i.e., $length(l) = 5$. Let the list l consist of five elements. We take the first element of the list and put it to the back of the list. Now, we do the same for the remaining four elements.

$$\begin{aligned} rotate(length([a,b,c,d,e]), [a,b,c,d,e]) &= rotate(5, [a,b,c,d,e]) \\ &= rotate(4, [b,c,d,e,a]) \\ &= rotate(3, [c,d,e,a,b]) \\ &= rotate(2, [d,e,a,b,c]) \\ &= rotate(1, [e,a,b,c,d]) \\ &= [a,b,c,d,e] \end{aligned}$$

It is very easy to see that this process gives us back the original list. Moreover, it is clear that if we follow the same procedure, i.e., schematic proof, for a list of any length, we always get back the original list. Hence, the number of inference steps in the proof depends on n, so the proof is schematic in n.

In contrast to a schematic proof of the *rotate-length* theorem, this theorem is not easy to prove by a conventional theorem prover. The inductive proof of the *rotate-length* theorem consists of a generalization: e.g., $rotate(length(l), append(l, k)) = append(k, l)$, where *append* is the list append function. It is harder to see that this theorem is correct. Schematic proofs avoid such generalizations. Baker used schematic proofs to exploit this fact for theorems of arithmetic (Baker et al. 1992).

We propose that the example given above demonstrates a common way that people think about the proof of this theorem. This proposal supports out conjecture that schematic proofs correspond better to human intuitive proofs.

4.6 Penrose, Gödel Argument and Constructive ω-Rule

We mentioned earlier in this book, that the work discussed in this book was partially inspired by Penrose's talk in 1995 to the Centre for Mathematical Sciences in Edinburgh. In this talk Penrose argued that the aim of the strong programme[24] in Artificial Intelligence (AI) is impossible. He claimed that there is something fundamentally non-computational in human mathematical reasoning, which therefore cannot be carried out by machines. There are several books in which Penrose argues his viewpoint on the difference between human mathematical reasoning and mathematical reasoning simulated on machines. See for example, Penrose 1989, 1994a,b.

Penrose uses an argument based on Gödel's first incompleteness theorem (see §B.1 and Gödel 1931) to argue for the non-algorithmic nature of human mathematical thought.[25] He argues that humans are capable by "insight" to see and check the correctness of any mathematical proof. By Gödel's first incompleteness theorem this is impossible for any formal system, and thus for machines:

> *"... it seems to me that it is a clear consequence of the Gödel argument that the concept of mathematical truth cannot be encapsulated in any formalistic scheme. Mathematical truth is something that goes beyond mere formalism."* Penrose 1989, page 145

The question arises whether the completeness of the system of Peano axioms with the constructive ω-rule disproves Penrose's argument. According to Penrose, the ω-rule and its constructive counterpart also suffer from the Gödel argument, namely that they are computationally infeasible. This is due to the infinitary nature of the rules. Although a complete formalization of arithmetic can be devised using Peano axioms and the constructive ω-rule, it turns out that the Gödel argument *does* apply to any meta-system in which the verification of the recursive function

---

[24]There is a spectrum of opinions about what the aim of strong AI is. We think that perhaps a generally accepted notion of strong AI is that we can create intelligence. Weak AI, on the other hand, argues that we can create behavior on machines which in humans would be considered to be intelligent.

[25]Gödel's first incompleteness theorem has been used in the past by Lucas, similarly to Penrose, to point out the distinction between reasoning by humans and reasoning by machines (Lucas 1970).

capturing the proof of a theorem is carried out.[26] Unfortunately, this appears to support Penrose's argument. However, our hypothesis is that humans often omit the inductive verification step. Curiously, Penrose himself omits this step in his argument (namely, in his presentation of the proof of the theorem about the *sum of hexagonal numbers*, explained below). In order to convince us that human mathematical reasoning is fundamentally non-computational and hence cannot be simulated on a machine, Penrose uses a procedure similar to the algorithm for implementation of the constructive ω-rule (given in §4.3).[27]

In his lecture in Edinburgh and in Penrose 1994a, Penrose gave an example of human mathematical reasoning in mathematical visualization, and claimed that such reasoning cannot be carried out by machines. The example that he used is the diagrammatic proof of a theorem about the *sum of hexagonal numbers.* This diagrammatic proof has been presented in §3.2.5. The theorem about the *sum of hexagonal numbers* states that the sum of first n hexagonal numbers is n cubed. In his proof Penrose (1994a) demonstrated only one instance of the proposition P, namely for $n = 6$. Thus, the sum of the first six hexagonal numbers (i.e., $1, 7, 19, 37, 61$ and 91) is six cubed (i.e., 216). He invited us to consider a cube of magnitude six and showed us how one can decompose this cube into six half-shells.[28] Each of these half-shells can be projected onto a plane to give a hexagonal number. Then, Penrose asked us to consider how general this procedure is, and that it would work for all values of n. To convince us, Penrose exhibited the trace for the proof for $n = 6$, i.e., $\mathsf{proof}(3)$, explained how to construct from this a general proof procedure proof, and claimed that $\mathsf{proof}(n)$ is correct, i.e., for each n it always gives a proof of the proposition $P(n)$.

Recall again the algorithm for using the constructive ω-rule in schematic proofs which was given in §4.3. Penrose's argument very closely follows this algorithm.

1. He proved one special case of the proposition. In particular, he gave $\mathsf{proof}(6)$ which is a proof of the proposition $P(6)$.
2. He discussed how to construct (abstract) a general proof procedure $\mathsf{proof}(n)$ from $\mathsf{proof}(6)$.
3. He claimed that $\mathsf{proof}(n)$ always proves $P(n)$.

[26] Recall that in §4.3 we gave a three-stage algorithm of how to apply the constructive ω-rule, and that the third stage was to verify in some meta-system that the recursive function which uniformly captures the proof is indeed correct.

[27] Most of the information about the line of argument that Penrose took at his talk in Edinburgh was communicated to me by Alan Bundy. Most of the analysis of Penrose's argument which is discussed in this section, is also due to Bundy.

[28] Recall that a half-shell consists of three adjacent faces of a cube. This terminology is not the one that Penrose used, but is due to Alan Bundy.

A careful consideration of Penrose's argument reveals that he is doing less than our algorithm in §4.3:

- He considers only one example of a proof.
- He does not formalize $\mathsf{proof}(n)$.
- He does not prove that $\mathsf{proof}(n)$ always proves $P(n)$.

Penrose's method for proving theorems is hence fallible. Potentially, a counterexample could be found, i.e., a value of n for which $\mathsf{proof}(n)$ does not prove $P(n)$. However, it seems that humans often use Penrose's method for solving problems. We, as human mathematicians, consider examples of proofs of a proposition and try to ensure that we consider all special cases and all types of examples. This corresponds to the first stage of Penrose's method. We then trust that our abstraction mechanism is general enough to encompass all the examples given in the first stage. This corresponds to the second stage of Penrose's method. Last, we rely on our judgment that the first two stages were carried out correctly, so we do not address the third stage of the method to check that our general proof is indeed correct. As mentioned in §4.5, this can be a possible explanation for the existence of erroneous proofs.

What is then an adequate automated proof checker, according to Penrose? An answer to this question bears importance in understanding Penrose's argument against strong AI. If we consider the three stages involved in his method of constructing proofs (and consequently in the implementation of the constructive ω-rule), then it seems that there should be no particular difficulty in automating each stage in a proof checker. Such a system would fulfill Penrose's challenge if the requirements are such as he uses in his own reasoning, namely they correspond to his own method of proof construction. In this book we present a system, called Diamond, which implements the procedure for construction of schematic proofs as given in §4.3, and therefore fulfills Penrose's challenge discussed in this section.

4.7 Diagrams and Schematic Proofs

We are now in the position to propose extending Baker's work on schematic proofs to our diagrammatic proofs so that the generality of the theorem and its proof is embedded in the schematic proof. Thus, we eliminate the need for abstraction devices in diagrams. A general schematic proof is constructed from geometric manipulations of concrete rather than general diagrams.

The notion of proof in formal logical theories is embedded in the application of rewrite rules. A theorem at hand is proved in a symbolic logical way (as opposed to a diagrammatic way) when the sentence expressing the theorem is reduced to a truth value, through an application of rewrite rules.

The notion of proof in diagrammatic reasoning of the kind that we presented in Chapter 3 is perhaps less obvious. The rewrite rules of a symbolic proof are replaced in a diagrammatic proof by geometric operations on a diagram. These could be seen as rewrite rules if they were part of some logical theory of diagrams.[29] The geometric operations transform a diagram in some way. Theorems that are part of our problem domain are theorems of natural numbers (see §3.5), therefore diagrams are represented using dots. All operations preserve the number of dots composing a diagram. In this way, we can appeal to the visual characteristic of the composition of dots (e.g., six consecutive rows of six dots form a square of magnitude six), and at the same time retain the notion of equality in the theorem (represented as an equation) throughout the application of geometric operations. The visual characteristic of the composition of dots gives us some sort of intuitive understanding of what a particular number represents (e.g., 6^2 is a square of magnitude six). The preservation of dots in a diagram convinces us that the operations are valid "diagrammatic rewrite rules" and that the equality is preserved. The notion of a diagrammatic proof is embedded in the transformation of diagrams representing one side of the equality (which states the theorem symbolically) into diagrams representing the other side of the equality. A diagrammatic proof is completed when one side of the equation is transformed into the other side of the equation (or equivalently, when the two sets of diagrams representing the two sides of the equation consist of identical diagrams). The notion of a proof, as implemented in our diagrammatic reasoning system DIAMOND, will be discussed in §5.3.

The process of a *diagrammatic* schematic proof starts with a few particular concrete cases of the theorem represented by diagrams. The diagrammatic manipulations (i.e., operations) on the diagram are performed next, capturing the inference steps of the diagrammatic proof. This step corresponds to the first step of the schematic proof procedure given in §4.3.

The second step is to abstract the operations involved to form a schematic proof for n. Note that the generality is represented as a recursive program which specifies a sequence of diagrammatic operations that are used on a diagram, and not as a general representation of a diagram. More precisely, the basic idea is to consider proofs for $n+1$ which

[29] For example, Hammer formalized a logical theory of Venn diagrams (Hammer 1995).

can be reduced to proofs for n (or conversely, such proofs for n which can be extended to proofs for $n+1$ by adding to them some additional sequence of operations). The difference between the proof for $(n+1)$ and the proof for n, i.e., the additional sequence of operations in the proof for $(n+1)$ with respect to the proof for n, is referred to as the step case of the abstracted schematic proof.

The last step in the schematic proof procedure is to prove by meta-induction that the abstracted diagrammatic schematic proof is correct. We need to show that $\mathsf{proof}(n)$ proves $P(n)$ for all n. One way of proving the correctness of schematic proofs is to create a theory of diagrams that models the processes in a diagrammatic reasoning system and prove correctness there. This will be discussed in Chapter 8.

We can see that the constructive ω-rule and schematic proofs can indeed be applied to diagrammatic theorems so that we can formalize diagrammatic schematic proofs. In the next section we give examples of schematic proofs for theorems of Category 2 which were presented in Chapter 3. The implementation of the formalization of diagrammatic schematic proofs and their construction will be the topic of Chapter 7.

4.8 Schematic Diagrammatic Proofs for Theorems of Category 2

We can now structure diagrammatic proofs in a more formal way. Identifying the geometric operations that are required to prove a theorem helps us define a sufficient repertoire of such operations[30] which are used in diagrammatic proofs. Diagrammatic proofs of Category 2 from §3.2 are structured here so that although the example-proofs were given for particular values of a parameter n, we present the proofs here in a general form. These general proofs are constructed by extracting a pattern from the trace of the example-proof procedure. In Chapter 7 we present how general schematic proofs are constructed automatically in Diamond. There are choices in the diagrammatic representation of (a part of) a theorem, which will be discussed in more detail in §5.3.1. For now it suffices to say that it is left to the constructor of the diagrammatic proof to make an informed choice of a diagrammatic representation of a theorem.

[30] By sufficient repertoire we mean a set of diagrams and operations which enable us to prove a significant range and depth of theorems. We discuss these issues in greater detail in §9.2.1.

4.8.1 Schematic Diagrammatic Proof for Sum of Odd Naturals

$$n^2 = 1 + 3 + \cdots + (2n - 1)$$

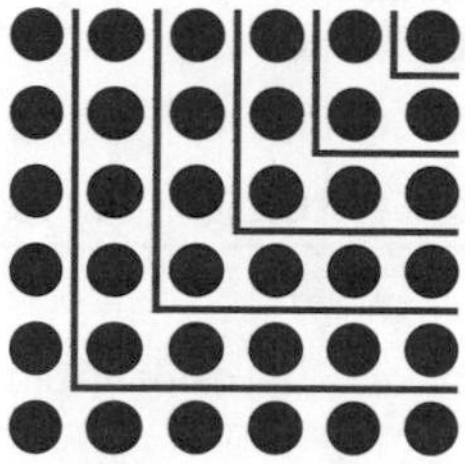

FIGURE 20 Sum of odd naturals.

Here we state the proof for the theorem about the *sum of odd naturals* (given in §3.2.3) as a sequence of steps that need to be performed on the diagram. Let us again choose that n^2 is represented by a square of magnitude n, $(2n-1)$ is represented as an ell and the natural number 1 is represented as a dot. The second step in the proof procedure justifies that an ell represents an odd number, i.e., that $2n - 1 = 2(n-1) + 1$:

1. Cut an ell from a square, where an ell consists of 2 adjacent sides of the square. This creates an ell of magnitude n and a square of magnitude $n - 1$.
2. For this ell, continue splitting from it pairs of dots at the end of two adjacent sides of the ell until only 1 dot is left (note that for each ell of magnitude n, we will have $n - 1$ pairs of dots plus another dot which is a vertex of the two adjacent sides, i.e., $2(n-1) + 1$).
3. Repeat these two steps on the remaining square until the square is exhausted.

Therefore, these steps are sufficient to transform a square of magnitude n representing the LHS of the theorem to n ells of increasing magnitudes representing the RHS of the theorem.

4.8.2 Schematic Diagrammatic Proof for Sum of Squares of Fibonacci Numbers

Here we give the proof for the theorem about the *sum of squares of Fibonacci numbers* (given in §3.2.4) as a sequence of steps that need to be performed on the diagram. Let $Fib(n+1) \times Fib(n)$ be represented by a rectangle of length $Fib(n+1)$ and height $Fib(n)$, and $Fib(n)^2$ by a square of magnitude $Fib(n)$:

$$Fib(n+1) \times Fib(n) = Fib(1)^2 + Fib(2)^2 + \cdots + Fib(n)^2$$

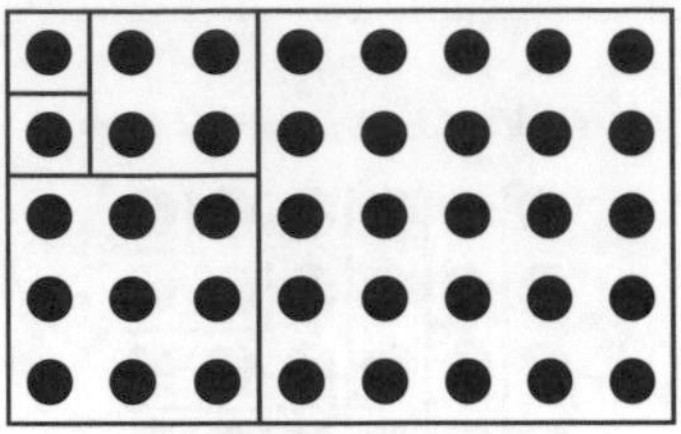

FIGURE 21 Sum of squares of Fibonacci numbers.

1. Split a square from a rectangle. The square should be of a magnitude that is equal to the smaller side of a rectangle (note that aligning squares of Fibonacci numbers in this way is a method of generating Fibonacci numbers, i.e., $1, 1, 1+1 = 2, 1+2 = 3, 2+3 = 5$, etc.).
2. Repeat this step on the remaining rectangle until it is exhausted.

Therefore, these steps are sufficient to transform a rectangle of magnitude $Fib(n+1)$ by $Fib(n)$ to a representation of the RHS of the theorem, i.e., n squares of magnitudes that are increasing Fibonacci numbers.

4.8.3 Schematic Diagrammatic Proof for Sum of Hexagonal Numbers

$$n^3 = Hex(1) + Hex(2) + \cdots + Hex(n)$$

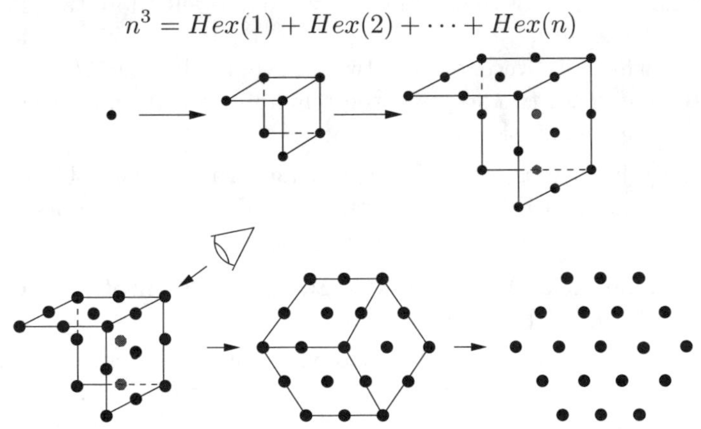

FIGURE 22 Sum of hexagonal numbers.

Here we state the proof for the theorem about the *sum of hexagonal numbers* (given in §3.2.5) as a sequence of steps that need to be performed on the diagram. Let n^3 be represented by a cube of magnitude n and $Hex(n)$ by an nth hexagon:

1. Split a half-shell from a cube (recall that a half-shell consists of three adjacent faces of a cube). This creates a half-shell of magnitude n and a cube of magnitude $n-1$.
2. Project this half-shell down the main diagonal of a cube from a three-dimensional space onto a plane (note that this forms a hexagon).
3. Repeat these two steps on the remaining cube until it is exhausted.

Therefore, these steps are sufficient to transform a cube of magnitude n representing the LHS of the theorem to n increasing hexagons representing the RHS of the theorem.

4.8.4 Schematic Diagrammatic Proof for Triangular Equality for Even Squares

$$(2n)^2 = 8Tri(n-1) + 4n$$

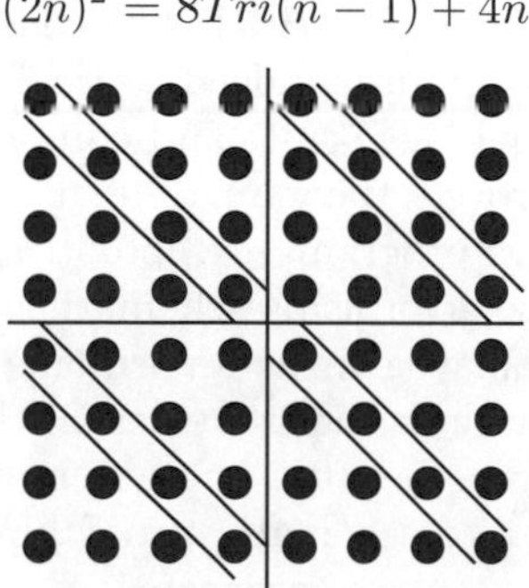

FIGURE 23 Triangular equality for even squares.

Here we list the proof for the theorem about the *triangular equality for even squares* (given in §3.2.6) as a sequence of steps that need to be performed on the diagram. The theorem is stated as $(2n)^2 = 8Tri(n-1)+4n$. Recall that the triangular numbers $Tri(x)$ were defined in §3.2.6. Let us choose to represent $(2n)^2$ with a square of magnitude $2n$ for some particular n. Note that there could be other diagrammatic representations of $(2n)^2$. Also, let us represent $Tri(n-1)$ as a triangle of magnitude $n-1$, and n as a side of magnitude n. The aim is then to transform a square of magnitude $2n$ into eight triangles of magnitude $n-1$ and four sides of magnitude n, for some particular n. A schematic proof is given for a general n:

1. Split a square of magnitude $2n$ into four identical squares (note that each of the four squares is of magnitude n).
2. Split each new square down the main diagonal (note that for each square, two triangles are created, one of magnitude n and one of magnitude $n-1$).
3. For each bigger triangle (of magnitude n), split from it one side (note that this creates another four triangles of magnitude $n-1$ and four sides of magnitude n).

Notice that the number of steps in this proof is constant, i.e., not dependent on the value of n. We call such proof trivially schematic.

These steps are sufficient to transform a square of magnitude $2n$ representing the LHS of the theorem to eight triangles of magnitude $n-1$ and four sides of magnitude n representing the RHS of the theorem.

4.9 Summary

A schematic proof is a recursive program which by instantiation at n gives a proof of every proposition $P(n)$. The constructive ω-rule justifies that such a recursive program is indeed a proof of a proposition for all n. The use of the constructive ω-rule can be implemented by providing a uniform procedure to prove a theorem. The uniformity of the procedure is captured in a recursive program, e.g., $\mathsf{proof}(n)$.

There are many reasons for using schematic proofs. Schematic proofs and their construction may give us an explanation for why there are erroneous "proofs" in mathematics that persist through many years. We speculate that this may be because humans use a procedure similar to the one for construction of schematic proofs, but they often omit the stage which verifies the recursive program.

Part of the inspiration for the research presented in this book came from Penrose and his talk in Edinburgh. We challenge Penrose's argument that human mathematical reasoning is fundamentally non-computational and thus that it cannot be automated. In this book we show how this can be done and make the first steps in doing so.

The constructive ω-rule and schematic proofs are particularly useful in our domain, since they provide a way of using concrete examples of proofs, and still allow one to conclude a general statement in the end. Hence, they can be used in diagrammatic reasoning. A diagrammatic schematic proof consists of an application of geometric operations on a diagram. Instead of using general diagrams which use abstraction devices, we capture the generality of a proof in a variable number of applications of geometric operations. This approach gives scope to a mechanism which can be implemented on machines – we show how this can be done next. Namely, we formalize the ideas presented so far and implement them in a diagrammatic reasoning system DIAMOND.

5

Designing a Diagrammatic Reasoning System

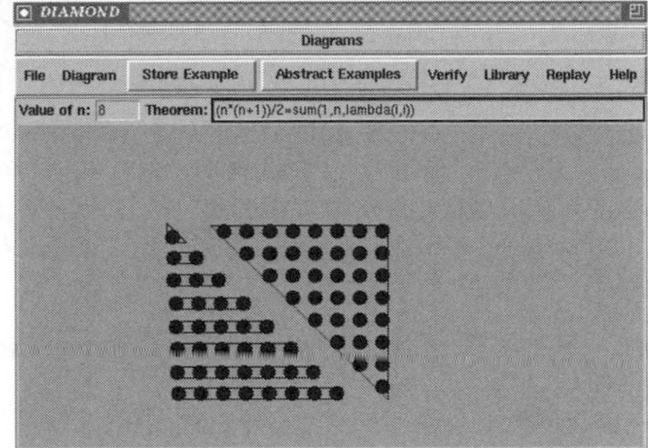

— DIAMOND
the System

In this book we want to demonstrate that diagrammatic reasoning can be automated and that diagrams can be used for formal proofs. To realize the formalization of diagrammatic reasoning we implemented a system called DIAMOND which proves theorems by using diagrammatic inference steps. DIAMOND is a diagrammatic proof checker, which interactively proves theorems of mathematics by applying geometric operations to diagrams. In this chapter some of the design issues for the implementation of this proof system are discussed.

In §5.1, a brief overview of DIAMOND is given. The architecture of DIAMOND is presented in §5.2. In §5.3 we describe the basic notion of a diagrammatic proof. In §5.4 we explain the construction of example-proofs. Finally, the discussion of other design issues for construction of proofs is given. They include the representation of diagrams in §5.5, and DIAMOND's graphical interface in §5.6.

5.1 Overview of DIAMOND

The DIAMOND system is an embodiment of the ideas presented in this book. DIAMOND stands for **Dia**gram**m**atic Reas**on**ing and **D**eduction. It is a diagrammatic proof system implemented in the functional programming language Standard ML version 109.[31]

In DIAMOND we capture the generality of diagrammatic reasoning by a diagrammatic schematic proof (see §4.3). The construction of a diagrammatic schematic proof in DIAMOND consists of three steps corresponding to the procedure described in §4.4.

- **The *interactive* construction of example-proofs.**
  This is the topic of this chapter. An example-proof is constructed interactively with the user. It consists of a sequence of geometric operations that are applied to a concrete diagram. The repertoire of geometric operations will be discussed in Chapter 6. This sequence in some way (explained in more detail in §5.3) justifies, i.e., proves, some ground instance of the theorem. In particular, if a theorem is expressed as an equality, then an instance of a proof transforms the diagrammatic representation of the left hand side of the equality through a sequence of number preserving operations into the diagrammatic representation of the right hand side of the equality.
- **The *automatic* construction of a schematic proof.**
  DIAMOND abstracts the concrete, interactively constructed example-proofs in order to construct a schematic proof that will hopefully be applicable to any ground instance. A schematic proof captures the generality using the variable number of applications of geometric operations to a diagram. This number of applications is some function of a parameter n, where n is a natural number. If two instantiations of a proof procedure have a common structure, then this structure is automatically abstracted by DIAMOND. The constructive ω-rule, introduced in Chapter 4, is used to justify that a general schematic proof does constitute a formal proof. The representation of schematic proofs and their automatic construction will be discussed in detail in Chapter 7.
- **The verification of a schematic proof.**
  The schematic proof is an abstraction of the example-proofs, and is an educated guess induced by the abstraction mechanism. It still needs to be formally verified that the schematic proof proves the theorem at hand. In particular, we need to show that for any

[31]For more information on the Standard ML programming language, see Paulson 1991.

instance n a schematic proof proof generates a correct proof of a proposition $P(n)$. One way of showing that proof(n) proves proposition $P(n)$ is to reason with general diagrams about proof(n). However, this re-introduces the need for abstraction devices in such diagrams. Alternatively, the correctness of proof(n) can also be shown by creating a theory of diagrams which models the processes in DIAMOND, and by carrying out a meta-level proof of correctness in this theory. The verification of schematic proofs will be discussed in detail in Chapter 8.

5.2 Architecture

DIAMOND consists of a *diagrammatic component* and an *inference engine*. The diagrammatic component forms and processes diagrams. The inference engine deals with the diagrammatic inference steps. It processes the operations on diagrams.

1) Inference engine: This is the main component of DIAMOND. It is the knowledge base component of the system. It consists of several parts or submodules:

- Assertion submodule: This accepts from the user a suggested diagram from which to start the diagrammatic example-proof.
- Operations submodule: This generates strings of constraints and geometric operations that are to be carried out on a diagram. It accepts from the user the diagrammatic operations to be used, and executes them. See Chapter 6 for detailed discussion of the operations in DIAMOND.
- Example-proof submodule: This keeps track of all the operations applied to a diagram. The operations and the states of diagrams are recorded in an execution trace referred to as an example-proof. This is discussed in §5.4.
- Abstraction submodule: This is the implementation of the abstraction mechanism which is used to construct schematic proofs from example-proofs. This abstraction fulfills the requirement of the constructive ω-rule for a uniform computable procedure (see §4.2). See Chapter 7 for detailed discussion of the abstraction method.
- Verification submodule: This checks that a schematic proof induced by the abstraction mechanism is indeed correct, i.e., that the schematic proof proves the proposition at hand. The verification is carried out in a theory of diagrams, which models the processes in DIAMOND. See Chapter 8 for detailed discussion of the verification mechanism.

- Import submodule: This accesses previous externally stored schematic proofs and adds them to the library of accessible proofs.
- Replay submodule: This instantiates diagrammatic proofs for a particular user-defined value of a parameter n. The effect is a simulation of an example-proof.

2) Diagrammatic component: this is the interface between the inference engine and the user. The Cartesian representation of the diagram is used in this component to draw diagrams on the screen. The effects of the operations that are applied to the diagram by the inference engine are shown here. The interface is presented in greater detail in §5.6.

5.3 DIAMOND's Notion of Proof

DIAMOND's notion of a proof is captured by a sequence of operations that need to be applied to an initial diagram. The initial diagram (or diagrams) that the geometric operations are applied to is a diagrammatic representation of the left hand side (LHS) of an instance of a symbolically expressed theorem. The result of applying all the operations of the diagrammatic proof to this diagram should be the diagrammatic representation of the right hand side (RHS) of the same instance of the symbolically expressed theorem.[32] A parametrized sequence of geometric operations for a particular theorem that fulfills this requirement for all instances of the parameter constitutes a diagrammatic proof.

DIAMOND is not a fully automated theorem prover. Rather, it is a proof checker. With DIAMOND we are not trying to discover diagrammatic proofs, but rather we are exploring and trying to understand them better. Thus, it is generally expected that the user has a diagrammatic proof in mind. However, if this is not the case, the user can simply try various combinations of diagrams and operations on them to explore their effects. It is up to the user to choose the appropriate diagrammatic representation of the symbolic theorem which is to be proved. For instance, the user would usually pick a square to represent n^2. There are choices that can be made and the user makes these choices according to the particular proofs that he or she has in mind.

5.3.1 Diagrammatic Representation of Arithmetic Expressions

A theorem of natural number arithmetic can have a diagrammatic proof if it is expressed using terms that can be mapped into a diagrammatic form. There are some obvious mappings that can be used. Table 1 gives

[32] Rewriting the LHS to get the RHS of the equation is a common technique in automated reasoning systems (Dershowitz and Jouannaud 1990).

some examples. Note that a diagrammatic representation is described for a particular value of n and m.

TABLE 1 Some diagrammatic representations of arithmetic expressions.

Arithmetic Expression	Diagram
n	row of magnitude n
n	column of magnitude n
n^2	square of magnitude n
$n \times m$	rectangle of magnitude n by m
n^3	cube of magnitude n

There are also some less obvious mappings from arithmetic expressions to diagrams. For example, one could choose two adjacent sides of a square to represent odd natural numbers. A triangle of magnitude n represents $\frac{n(n+1)}{2}$, since the domain of theorems is natural number arithmetic (as opposed to $\frac{n^2}{2}$ for any real number n).[33] The circumference (also called the frame) of a square of magnitude n where n is a natural number represents $n^2 - (n-2)^2$ or, equivalently, $4(n-1)$. These mappings do not necessarily need to be obvious to a human. Rather, they can be constructed in a way which would make the equivalence explicit in order for a human to understand what arithmetic expression they represent. For instance, to explain that a frame of a square of magnitude n represents $n^2 - (n-2)^2$ one just has to imagine taking a square of magnitude n and remove from it an inner square which is of magnitude $n-2$. On the other hand, if two rows and two columns of magnitude $n-1$ are joined at the sides they form a frame, hence $4(n-1)$.

Clearly, the domain of theorems which can be proved in a diagrammatic way is restricted by the possible diagrammatic representations of a theorem. A set of given diagrams and operations on them can be used to construct the less explicit mappings of arithmetic expressions into diagrams. The choice of which diagram represents which expression depends on the particular example-proof that the user has in mind. Some choices are better than others, because an appropriate diagram enables

[33]In continuous space one can think of an area of a right angle triangle of magnitude n as half of a square of magnitude n if this square is split along its diagonal, hence $\frac{n^2}{2}$. However, in discrete space a square is represented using dots. Splitting a square along its diagonal does not split it to two identical triangles, because the corner dots on each side of the diagonal cannot be halved. Hence this creates one triangle of magnitude n and one of magnitude $n+1$. Taking a rectangle of magnitude n by $n+1$ and halving it creates two triangles of magnitude n, hence a triangle represents $\frac{n(n+1)}{2}$.

the use of appropriate operations on the diagram which are necessary to carry out the proof. For instance, consider the theorem about the *sum of hexagonal numbers* given in §3.2.5. The theorem states that the sum of n hexagonal numbers is equal to the cube of n. The diagrammatic proof given in §3.2.5 consists of splitting half-shells[34] from a cube of magnitude n, plus some additional operations. Were we to choose that n^3 is represented diagrammatically as n squares of magnitude n for some particular natural number n, then the proof could not be carried out, because the operation of splitting a half-shell from a cube would not be possible. The selection of the diagrammatic representation of an arithmetic expression in a diagrammatic proof corresponds to the choice of mathematical induction or a lemma in a symbolic proof of a theorem. If the appropriate representation (in a diagrammatic proof) or induction scheme (in a symbolic proof) is selected, then the proof can be carried out. This will be discussed in greater detail when multiple representations of diagrams and operations on them are introduced in §6.5.

The choices for the mapping of arithmetic expressions into diagrams could in principle be automated – but this is a topic for future work (see §10.1).

5.4 Construction of Example-Proofs

DIAMOND's example-proofs consist of a sequence of applications of geometric operations to a diagram. The operations are the inference steps of the proof. Example-proofs are interactively constructed for particular concrete values of a parameter n. DIAMOND records a trace of the operations used in each example-proof. The idea is to compare example-proofs and detect if there is a common structure between them. If so, then we want to capture this common structure in a general way. We try to find a proof such that $\mathsf{proof}(n)$ proves a proposition $P(n)$ for all n. So, for example, consider two instances i_1 and i_2 of a universally quantified variable n. Also, let $example\_proof_1$ and $example\_proof_2$ be two example-proof traces for i_1 and i_2. Then, we at least require that:

$$\begin{aligned} \mathsf{proof}(i_1) &= example\_proof_1 \\ \mathsf{proof}(i_2) &= example\_proof_2 \end{aligned}$$

The aim is to find a uniform and effective characterization of proof to capture the generality of the proof, i.e., a recursive function proof parametrized over n. We employ heuristics in automating the construction of such a function. The formalization and construction of the recursive program capturing a general proof will be described in Chapter 7.

[34]A half-shell is a combination of three adjacent faces of a cube.

A particular formalization of a recursive program depends on the structure of the example-proofs. For instance, assume that an example-proof for some $n+c$ (where $c \neq 0$, and n and c are some natural numbers) can be constructed using an example-proof for this n plus some additional operations, e.g.:

$$\begin{aligned} \mathit{example\_proof}(n+c) \quad = \quad & \mathit{operations}_1(n+c) \text{ then} \\ & \mathit{operations}_1(n) \text{ then} \\ & \vdots \\ & \mathit{operations}_2 \end{aligned}$$

Then the recursive program which represents this example-proof can be formalized with only one recursive call to itself, e.g.:

$$\begin{aligned} \mathsf{proof}(n+c) &= \mathit{operations}_1(n+c) \text{ then } \mathsf{proof}(n) \\ \mathsf{proof}(c) &= \mathit{operations}_2 \end{aligned}$$

On the other hand, if for instance, an example-proof for a particular $n+c$ consists of operations which can be reorganized as follows:

$$\begin{aligned} \mathit{example\_proof}(n+c) \quad = \quad & \mathit{operations}_1(n+c) \text{ then} \\ & \mathit{operations}_1(n) \text{ then} \\ & \vdots \\ & \mathit{operations}_2 \text{ then} \\ & \mathit{operations}_1(n) \text{ then} \\ & \vdots \\ & \mathit{operations}_2 \end{aligned}$$

then the formalization of the recursive program can be:

$$\begin{aligned} \mathsf{proof}(n+c) &= \mathit{operations}_1(n+c) \text{ then } \mathsf{proof}(n) \text{ then } \mathsf{proof}(n) \\ \mathsf{proof}(c) &= \mathit{operations}_2 \end{aligned}$$

where there are two recursive calls in the program.

Each of the different recursive programs gives a different proof. In DIAMOND we are interested in proofs of inductive theorems. Therefore, DIAMOND expects the example-proofs to be formulated in a particular way where the order of operations is important. Namely, example-proofs are expected to be given with the same order of operations, perhaps with some extra operations in the case of the proof for $n+c$ with respect to the proof for n for some particular n. There is some justification of this constraint on the order of operations: it follows an inductive argument where instances of theorems for $n+c$ can be proved using proofs of instances of theorems for n, which can be proved using proofs of instances of theorems for $n-c$ etc. We say that this class of theorems is characterized by the smallest value of c. However, the user is not constrained to provide example-proofs for two consecutive values n and $n+c$. On the contrary, the user is allowed to provide any two examples of the same class (with characterizing c), i.e., for n and $n+kc$ for any $k \neq 0$. The importance of the order of operations is due to the limitation of the abstraction mechanism (see Chapter 7). If the example-proofs do not satisfy this constraint, the abstraction technique cannot detect the common structure – this limitation requires improvement in the future.

Consider the example for the *sum of odd naturals*. The theorem is symbolically stated as $n^2 = 1 + 3 + 5 + \cdots + (2n - 1)$. The user can choose a square amongst the available diagrams to represent n^2 on the left hand side of the theorem. The user can also choose operations such as splitting two adjacent sides from a square, and splitting the ends from these two adjacent sides. These are the operations that will be used in the example-proof presented here. The example-proof is given for concrete values. Take $n = 4$ and the instance $4^2 = 1+3+5+7$, and $n = 3$ and the instance $3^2 = 1 + 3 + 5$. Figure 24 shows the interactively constructed

1. Cut a square 4 times into ells, where an ell consists of 2 adjacent sides of the square.

2. For each ell, split end dots from both edges $(n - 1)$ times (i.e., $3, 2, 1$ and 0 times).

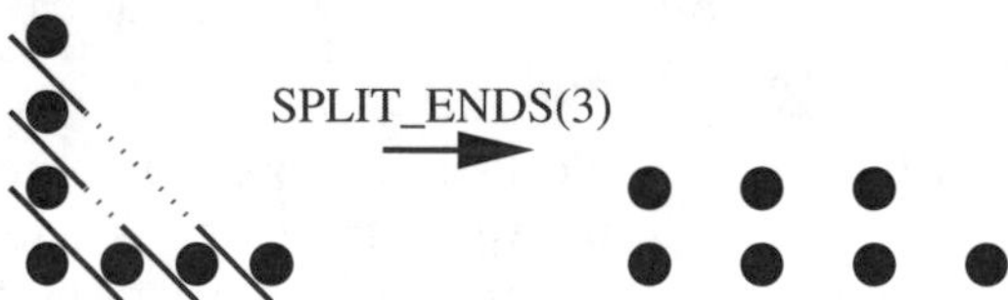

FIGURE 24 Sum of odd naturals for $n = 4$.

example-proof for $n = 4$ and Figure 25 shows another example-proof for $n = 3$.

The first part of these example-proofs decomposes a square into ells: in the case of $n = 4$ into four ells, and in the case of $n = 3$ into three ells.

1. Cut a square 3 times into ells, where an ell consists of 2 adjacent sides of the square.

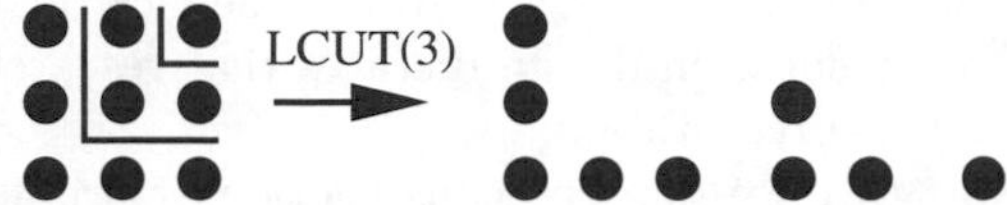

2. For each ell, split end dots from both edges $(n-1)$ times (i.e., 2, 1 and 0 times).

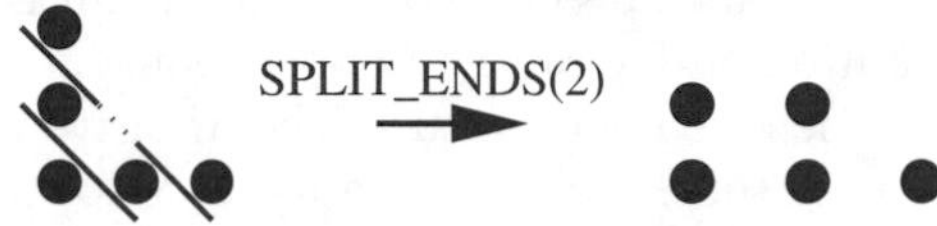

FIGURE 25 Sum of odd naturals for $n = 3$.

This corresponds to the number of elements summed in the instantiated theorems. The second part shows that each ell represents an odd natural number, which corresponds to $2i - 1$ for each i in the sum in the instantiated theorems.[35] The execution trace for the example-proof where $n = 4$ that DIAMOND records consists of the following operations: [lcut, split_ends, split_ends, split_ends, lcut, split_ends, split_ends, lcut, split_ends, lcut].

After the user generates two example-proofs, DIAMOND needs to abstract them and construct a general schematic proof from them. The representation and the construction of schematic proofs will be presented and discussed in Chapter 7.

5.5 Representations

One of the important realizations of mathematical reasoning research has been that the representation of knowledge is critical to one's ability to find the solution to the problem. It was Pólya who was first to advise us on the importance of knowledge representation (Pólya 1945). Simon argued Pólya's point further in Simon 1996 by stating that solving a problem means representing it so that the solution becomes trivial, or at least transparent. Amarel (1968) was the first one to consider this problem more closely in the context of *automated reasoning*. There has been much research done in the area of the automation of the representation design, but unfortunately not much achieved (Kulpa 1994). However, a promising approach has been taken by Van Baalen (1989). He

[35] If it is already known that an ell represents an odd number, then the second part of the proof can be omitted.

proposed an automated representation design method. Unfortunately, Van Baalen's approach is targeted for predicate calculus representation, rather than diagrammatic representation. The lack of success in the automation of representation design dictates that researchers devise their own appropriate representation.

In the DIAMOND system the construction of proofs is entirely diagrammatic, i.e., we use only diagrams for inferencing. Thus the knowledge representation needs to be diagrammatic as well. In particular, one of the design issues which needs to be considered in DIAMOND is the *internal* representation of diagrams and operations on them. We choose a representation which we hope captures the intuitiveness, rigor and simplicity of human reasoning with diagrams. DIAMOND is able to inspect and manipulate diagrams in a way that does not allow unsound inferences. Moreover, the operations on a diagram are easily carried out by the system. The external representations of diagrams via the user interface are simple and hopefully comprehensible to any user. Considering the advice of Pólya and Simon about the importance of representation, we built in DIAMOND a problem representation which enables it to obtain a solution by an automated construction of diagrammatic proofs. With their visual perception humans can observe and inspect diagrams directly and see (depending on how accustomed we are to mental spatial manipulations) the inference that needs to be made to prove a theorem diagrammatically. We capture some of the simplicity of human visual perception, and represent diagrams in a way which enables a theorem prover to prove theorems using diagrammatic inference steps.

In our choice of representation for achieving this, we considered several candidates, including:

- Cartesian representation – see recent translation Descartes 1954,
- Projective geometry – see Zisserman 1992,
- Diagrams on a raster – see Furnas 1990,
- Vector representation – see Larkin and Simon 1987,
- Topological (relational) representation – there is no one definitive reference for this, but see some overview of representations in Narayanan 1992, Kulpa 1994, Sloman 1996.

In DIAMOND we chose to use a mixture of Cartesian and topological representations. In the next few sections we analyze the use of these two representations with respect to the requirements in the implementation of DIAMOND. We justify our choice of mixed representation and show how diagrams are represented internally in DIAMOND. For more information on the other representations, the reader is referred to the literature cited above.

5.5.1 Cartesian Representation – Why Not Alone?

This is a commonly used representation in geometry. Examples of systems which use Cartesian coordinates for internal representation of diagrams are the Geometry Machine (see §2.1 and Gelernter 1963) and Polya (McDougal and Hammond 1993). Diagrams are represented in terms of the coordinate system, typically two or three dimensional. In a two dimensional space a point is a pair of numbers which is the coordinate. Diagrams can be represented as lists of coordinates. For instance, a possible Cartesian representation of a cube is as follows:

$$\begin{aligned} Cube \quad = \quad & \{(0,0,0),(1,0,0),(0,1,0),(1,1,0),(0,0,1), \\ & (1,0,1),(0,1,1),(1,1,1,)\} \end{aligned}$$

There should also be some indication of which points are joined together by lines, and which lines form faces. Carrying out operations on geometric objects represented by Cartesian coordinates requires matrix or other kinds of symbolic manipulations. These manipulations can be complex and unintuitive even for simple operations. However, computers are efficient at symbolic manipulations of diagrams represented by Cartesian coordinates. When proving conjectures symbolically, the complexity of matrix procedures required to represent the geometric manipulations on objects is not a problem. Furthermore, in systems that reason symbolically the unintuitiveness of matrix manipulations also does not seem to be a problem, because the system can still reason efficiently.

On the other hand, in DIAMOND we do not construct proofs with symbolic manipulations of diagrams. Moreover, DIAMOND's operations should be intuitive and easily carried out. Take, for instance, a geometric operation that might be needed in DIAMOND: an operation which splits a face from a cube. First, the system needs to distinguish which of the six faces is to be split from a cube. When this is established, the operation can be carried out. The result of the operation is two cuboids. Using the Cartesian representation of a cube, it is difficult to see which coordinates represent this particular face of the cube, and which coordinates represent the rest of the cube. In DIAMOND we want such an operation to be readily carried out, whereby the user points to the face of the cube and the system splits the face from a cube. Hence, when using DIAMOND the user does not need to worry about any complex matrix manipulations which are otherwise required if using Cartesian representation. It seems therefore, that Cartesian representation alone is not appropriate for the internal representation of diagrams in DIAMOND.

5.5.2 Topological Representation – Why Not Alone?

Topological (also called relational) representation is, in contrast to Cartesian representation, independent of any coordinates. It expresses the relations between the elements of the diagram. For instance, if we have a square *abcd*, then its topological representation might look like this:[36]

point(a)	segment(ab)	angle(abc)
point(b)	segment(bc)	angle(bcd)
point(c)	segment(cd)	angle(cda)
point(d)	segment(da)	angle(dab)
	segment(XY)=segment(YX)	angle(XYZ)=angle(ZYX)
	segment(XY)=segment(WZ)	angle(XYZ)=angle(QPR)

Topological representations are easy to implement on computers. They can vary in the degree of detail explicitly represented. For instance, if the information about the angles of a square is not needed to solve a problem, then such information does not need to be specifically stated. Grover (see §2.3 and Barker-Plummer and Bailin 1992) is one of the systems that uses topological representation for internal representation of diagrams. The downside is that topological representation can be too specialized for some problems, especially when numerical information about a diagram is required to solve a problem.

The problem with the topological description of a square as given above is that it is a very specialized one, particularly suited for problems that deal with relational characteristics of a diagram. On the other hand, in DIAMOND we are not interested in the fact that some angles in a diagram are equal to some others, for example. As in the human visual perception of angles, we want this fact to be transparent in the representation of a diagram. Consider again one of the operations that we might want in DIAMOND: to split a square along its diagonal. It is not easy to see which parts of the square representation given above will represent one resulting triangle and which will represent the other resulting triangle. It seems therefore, that topological representation alone is not appropriate for the internal representation of diagrams in DIAMOND.

5.5.3 Mixed Representation

Consider the problem domain (presented in Chapter 3) that DIAMOND is targeted for: theorems of natural number arithmetic. Diagrams represent natural numbers, so the representation of diagrams and operations on them should reflect the effect that operations have on diagrams with respect to the natural numbers that particular diagrams represent. Con-

[36]Note that the lower-case letters are used for constants and the upper-case letters are used for variables.

sidering the taxonomy of diagrammatic theorems given in §3.3, in particular, theorems of Category 2, suggests that in DIAMOND we are not interested in geometric properties of diagrams (such as the magnitudes of angles or which segments are parallel to each other). Rather, we are interested in the effect of splitting parts of diagrams apart in particular ways, and the effect of the operations on the natural numbers that the diagrams represent.

The representation of diagrams should be pertinent to the operations on them, so that the operations can easily be carried out. For instance, were we to split a face from a cube, then one of the good representations of a cube could be in terms of a sequence of faces comprising a cube. Furthermore, were we to split a square along its diagonal, a good representation of a square is in terms of two triangles.

It appears now that neither Cartesian nor topological representation alone meets these requirements. Topological representation alone can represent a square consisting of two triangles, but it does not specify how they are combined to form a square. Cartesian representation alone specifies the position of the square, but a complex matrix manipulation is required to split this square into, say, two triangles. Therefore, we decided that in DIAMOND a mixture of Cartesian and topological representation should be used for the representation of diagrams. First, we introduce DIAMOND's mixed representation, and then we explain why combining the two representations does not combine their individual disadvantages, but rather solves them.

It is essential to realize that in DIAMOND we need to represent only concrete diagrams, i.e., the ones that are of a particular magnitude. The magnitude of DIAMOND's diagram is always a natural number. We do not need to represent general diagrams with abstraction devices, since the generality of the proof is captured by a schematic proof (see Chapter 7).

The primitive object of DIAMOND is a `dot`, which represents the natural number 1.[37] This primitive object `dot` carries the information about the Cartesian coordinates. Thus we have `dot(x,y)` in the two dimensional space, and `dot(x,y,z)` in the three dimensional space, where `x`, `y` and `z` are instantiated to concrete natural number values. We shall refer to the primitive object as a ●. All DIAMOND's diagrams are either constructed from `dots` or from other diagrams. Examples of diagrams constructed of `dots` only include `row`, `column`, `ell` and `frame`. Examples of diagrams constructed from other diagrams include a square which is represented in terms of two triangles. Such representation ren-

[37]Were we to extend DIAMOND to a continuous space, we might want to consider a line or an area to be a primitive object.

ders splitting a square along its diagonal almost trivial. Examples of derived objects include `square, rectangle, triangle,...`

TABLE 2 Internal and external representation of a row and a square of magnitude 4.

Internal Representation	External Representation
`row (dot,dot,dot,dot)`	● ● ● ●

`square( row(dot,dot,dot,dot),`
`row(dot,dot,dot,dot),`
`row(dot,dot,dot,dot),`
`row(dot,dot,dot,dot))`
	● ● ● ●
	● ● ● ●
	● ● ● ●
	● ● ● ●

Table 2 shows the internal representation of some diagrams (a row and a square), and their external representation, as they appear in the user interface, and as humans may think of them (considering that the space is discrete).

Diagrams also have multiple representations. For instance, a square can be represented as a collection of rows, or as a collection of columns, or as a collection of ells, etc. This will be discussed in more detail in Chapter 6 where the operations are presented.[38]

The question now is why combining two different types of representation does not combine their problems, or introduce a new one. The reason is that the good sides of one representation remove the bad sides of the other, and vice versa. Hence, only the advantages of both representations remain. In particular, using topological representation solves the problem of complexity and unintuitiveness of Cartesian representation. For instance, a square can be represented as two triangles using the topological representation. This makes it easy to split a square into two triangles. Since each triangle is represented using dots with Cartesian coordinates, this makes it easy to see how these triangles are combined together. Moreover, it removes the need to specify the relations between, for instance, different angles, and other specialized properties of diagrams, because these are now implicit in the Cartesian representation.

[38]Multiple representations are presented alongside the description of operations due to a close relationship and interdependence between the representations of diagrams and operations on them.

5.6 Interface

The graphical interface of DIAMOND was implemented in SmlTk,[39] which is a Standard ML package providing a portable, typed and abstract interface to the user interface description and command language Tcl/Tk.[40] It allows the implementation of graphical user interfaces in a structured and reusable way, supported by the powerful module system of Standard ML. Figure 26 and Figure 27 show a screen shot of a DIAMOND session. There are three windows that are fired up when a DIAMOND

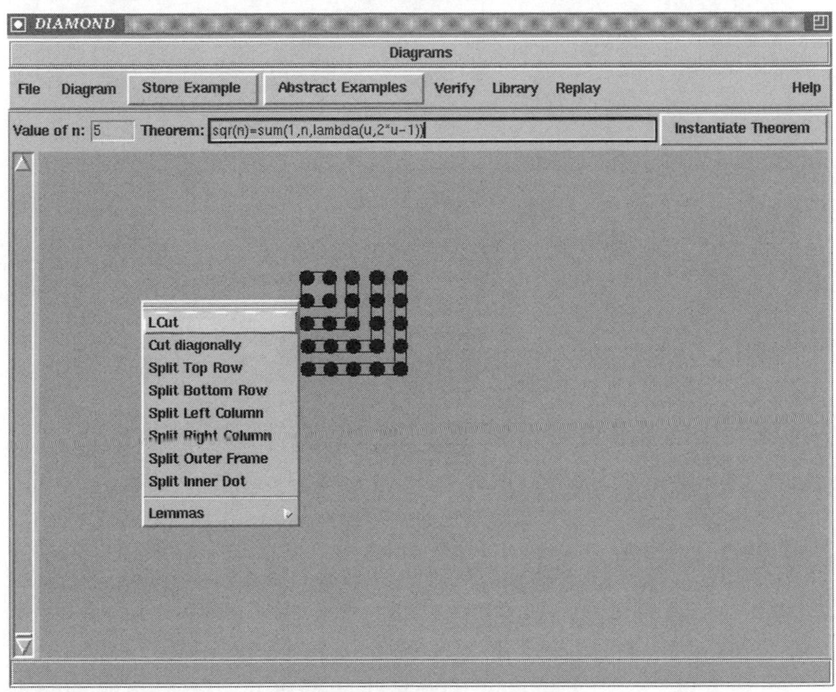

FIGURE 26 Screen shot of DIAMOND– main.

session is started. They are entitled *DIAMOND – Diagrams* shown in Figure 26, *PROOF TRACE* shown in Figure 27, and *clam-server* not shown here. The main window where the geometric operations are applied to a diagram is the *DIAMOND* window. It consists of a canvas

[39]SmlTk was implemented by C. Lüth, S. Westmeier and B. Wolff at the University of Bremen, Germany. It is publicly available via the Internet on the following site: `http://www.informatik.uni-bremen.de/~cxl/sml_tk/`.

[40]For more information on Tcl/Tk the reader is referred to Welch 1995.

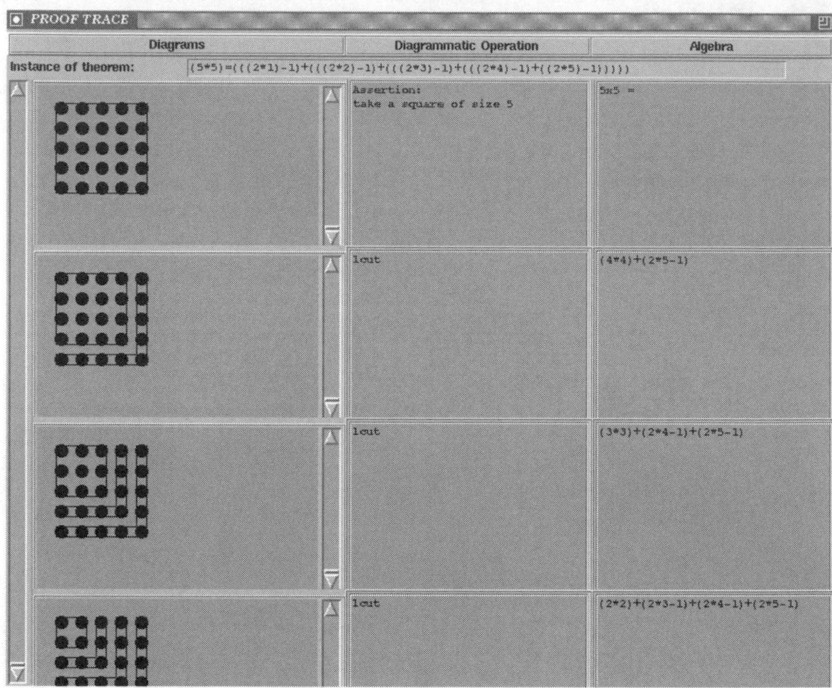

FIGURE 27 Screen shot of DIAMOND– trace.

where the diagrams are displayed, the field where the value for the particular parameter n (for which the example-proof is given) is entered, the field where the theorem at hand is entered, the `Instantiate Theorem` button which instantiates the entered theorem for the entered value of n, and a menu. The diagram menu consists of the following options:

File: this enables importing of previously saved schematic proofs (regardless of whether they have been verified or not), saving schematic proofs, starting new example-proofs and quitting the DIAMOND session.

Diagram: this enables the user to choose diagrams used in example-proofs – square, triangle, rectangle, ell, ...

Store Example: this is a button which enables the user to store example-proofs from which a schematic proof is constructed.

Abstract Examples: this is a button which executes the abstraction command, whereby DIAMOND automatically abstracts from example-proofs and constructs a schematic proof parametrized over some n. As DIAMOND successfully constructs a schematic proof, an additional option is added to the `Save` option in the `File` menu option, which enables the user to save the current schematic proof.

Verify: this is initially empty. When the user imports a file containing a previously saved schematic proof which was not checked for its correctness, or when a schematic proof is successfully abstracted during the current DIAMOND session, then the option of checking its correctness is added to this menu. It executes the verification command which checks if a schematic proof is correct. The window entitled *clam-server* displays the process of automatic verification.

Library: this is initially empty. When the user imports files containing previously saved schematic proofs, they are added to the library. The user can then choose to browse through the library of schematic proofs, and use an existing proof within another schematic proof as a submodule.

Replay: this is also initially empty. As schematic proofs are added to the library, they can be instantiated for a particular value of n, and then simulated (replayed) on the screen.

Operations on diagrams are accessed by clicking on the diagram (on the canvas) on which the operation is to be carried out so that the pop-up menu is activated. These pop-up menus are generated dynamically. They only enable a choice from those operations that are possible on the diagram. For instance, split_frame is an operation that is allowed on a square only. So, only the pop-up menu for a square will enable the use of this operation. The various operations and the diagrams on which they can be used will be discussed in detail in Chapter 6.

The second window entitled *PROOF TRACE* keeps track of the diagrams which are created at each step of the proof, and the operations which are applied to diagrams. It consists of three columns entitled *Diagrams*, *Diagrammatic Operations* and *Algebra*, and a field *Instance of Theorem* where the instance of the entered theorem for some entered value n is displayed. The *Diagrams* column displays the set of diagrams which are present at each step of the proof. The *Diagrammatic Operations* displays the diagrammatic operation which is applied to the diagrams at each corresponding proof step. Finally, the *Algebra* column displays the symbolic effect of each corresponding diagrammatic operation for a particular value of n. For instance, if we take a square of magnitude 5, the symbolic equivalent is 5^2. If we apply lcut once, this changes 5^2 to $4^2 + (2 \times 5 - 1)$. This will be discussed in more detail in §6.5. Note that we display symbolic terms equivalent to diagrams in order to convey better the notion of a proof. This symbolic representation of diagrams plays no role in the construction of diagrammatic proofs.

The third window entitled *clam-server* is just a text box which displays the verification process when the correctness of schematic proofs is checked. Verification of schematic proofs will be discussed in greater detail in Chapter 8.

5.7 Summary

We implemented the ideas presented in this book in a system called DIAMOND. DIAMOND is a diagrammatic proof checker which interactively proves theorems of arithmetic. DIAMOND's procedure to construct diagrammatic proofs consists of three steps: the interactive construction of example-proofs, the automatic construction of schematic proofs from example-proofs, and the verification of schematic proofs. This chapter dealt with the first part: the construction of example-proofs. The other two parts will be presented and discussed in subsequent chapters.

DIAMOND reasons with diagrams, hence the internal representation of diagrams is clearly important. Analysis of several techniques identified a suitable representation: a mixture of Cartesian and topological representation. This allows DIAMOND to specify only the relevant spatial relations in a diagram, and at the same times removes the need for complex symbolic manipulations. DIAMOND's architecture consists of an inference engine, which does the reasoning, and a diagrammatic component, which processes the diagrams. DIAMOND is an interactive system, so a usable graphical user interface is essential and has been implemented.

6

Diagrammatic Operations

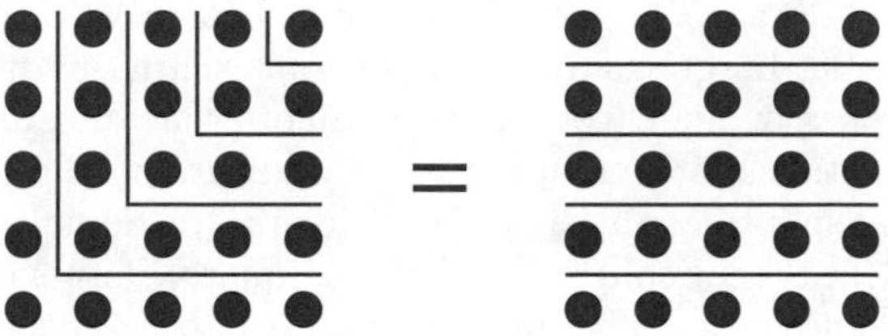

$$\sum_{i=0}^{n}(2i-1) \qquad = \qquad \sum_{i=0}^{n} n$$

— MJ

A diagrammatic proof, constructed by the system DIAMOND, consists of geometric operations that are applied to a diagram. In order to formalize diagrammatic proofs, the operations need to be formalized first.

This chapter presents the geometric operations which are available in DIAMOND. The availability of an operation depends on the representation of a diagram. Hence, we introduce multiple representations of diagrams in §6.2. There is a close relation between the representation of diagrams and operations on them – this is discussed next in §6.3. In §6.4 we give the analysis of the use of operations in tactics. Finally, in §6.5 the correspondence between the choice of an operation on a diagram (and consequently, the representation of a diagram), and the choice of an induction schema in a symbolic proof is demonstrated.

6.1 Choice of Operations

Operations are also referred to as manipulations or procedures. They capture the inference steps of DIAMOND's diagrammatic proof. Therefore, a fairly large number of such operations which are available to the user in the search for the proof, is identified and formalized. The intention is that the set of available operations enables one to prove theorems of significant range and depth. The justification of a significant range and depth is informal and is discussed in more detail in §9.2.1. To date, Diamond has been used to prove about thirty diagrammatic theorems. Extending the set of available diagrams and operations will enable Diamond to prove more theorems. The theorems range from non-inductive to inductive theorems. Nelsen's books *Proofs Without Words* (Nelsen 1993) and *Proofs Without Words II* (Nelsen 2001) have been used as the main source of examples.

In Chapter 5 some of the kinds of operations that are needed in Diamond were described in order to choose an appropriate representation for diagrams. To recap, Diamond is targeted to prove theorems of discrete arithmetic. Diagrams are a way of representing natural numbers. The interest lies in the effect on the numbers that diagrams represent after an operation has been applied to the diagrams. Thus, the operations join and split diagrams apart in various ways. Some operations are just simple ones (e.g., split a row from a square), and some are more complicated ones (e.g., decompose a square into a sequence of rows), which will be discussed in detail in this chapter.

There is a close relation between the choice of the diagram representation and the possible operations on this diagram. Ideally, the representation of the diagram is pertinent to the operation that is being carried out on it. Such a representation renders an operation very easy to apply. It is just a simple decomposition of the representation of a diagram. Different representations of a diagram allow different operations on it. We identify some of the possible representations of diagrams next.

6.2 Multiple Representations of Diagrams

The importance of problem representation has been acknowledged by many researchers (Simon 1996, Amarel 1968, Van Baalen 1989). Amongst them is George Pólya who argues in his books *"How to Solve It"* (Pólya 1945) and *"Mathematical Discovery"* (Pólya 1965) that the choice of representation of a problem is vital for finding its solution. In automated reasoning systems it is difficult to see how to use this advice, since there is normally only one representation scheme for the problem which is available to the system. An example of a commonly used representation scheme in automated reasoning system is predicate logic (Bundy 1983).

In Diamond, however, Pólya's advice of using alternative representations can be readily considered. Namely, diagrams can be represented in a variety of ways, and hence, theorems can have many different representations.[41] For instance, a square can be represented using several different compositions:

- a sequence of rows,
- a sequence of columns,
- a concentric sequence of circumferences, each of which is called a frame,
- a nested sequence of ells,
- a sequence of four squares, each of which is half the magnitude of the big one (note that the big original square has to be of even magnitude, and that the representation is recursive if the magnitude of the square if a power of 2),
- a matrix of dots,
- a sequence of diagonals.

Figure 28 shows these possible representations.

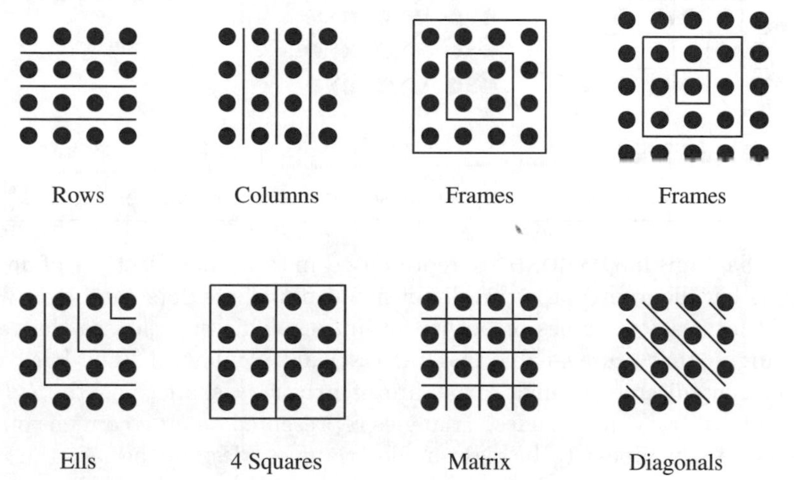

FIGURE 28 Multiple representations of a square.

Some of the multiple representations of a rectangle are analogous to the ones of a square, some are not applicable, and some are new. A rectangle can be represented as follows:

[41] The reader may notice that for some cases there is a connection between representations of a diagram and induction schemas. This connection will be discussed in greater detail in §6.5.

- a sequence of rows,
- a sequence of columns,
- a nested sequence of squares,
- a matrix,
- a sequence of diagonals.

Figure 29 shows these possible representations.

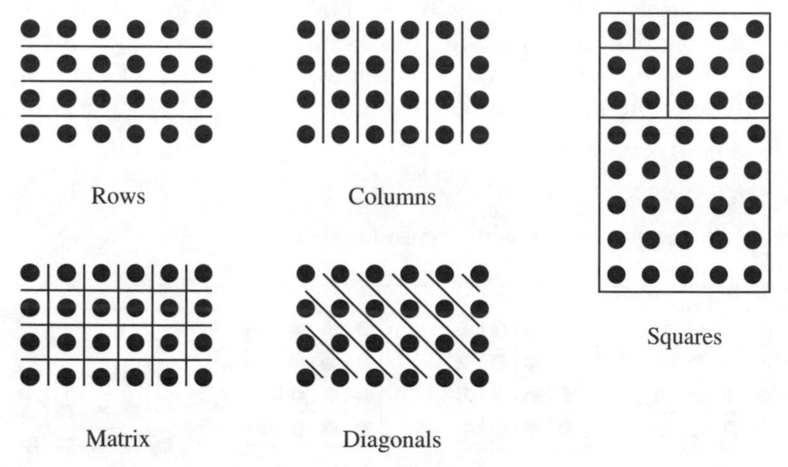

FIGURE 29 Multiple representations of a rectangle.

Diagrams in DIAMOND are represented in terms of collections of `dots` (or other diagrams) on a two-dimensional net where dots can be drawn only for discrete values of both coordinates. This necessitates that all triangles that are available in DIAMOND are equilateral. It is hard to represent discrete triangles that are of arbitrary shape, i.e., the sides are of arbitrary magnitudes. Triangles represented on a two dimensional discrete net appear to be right-angle triangles, despite the fact the all the sides of any triangle are of equal discrete magnitude. If DIAMOND was extended to prove theorems of real arithmetic, then there would be a need for a continuous space, and therefore scope for triangles of any magnitude. A triangle in DIAMOND can be represented as:

- a nested sequence of sides,
- a nested sequence of ells,
- a collection of two triangles and a square.

Figure 30 shows these possible representations.

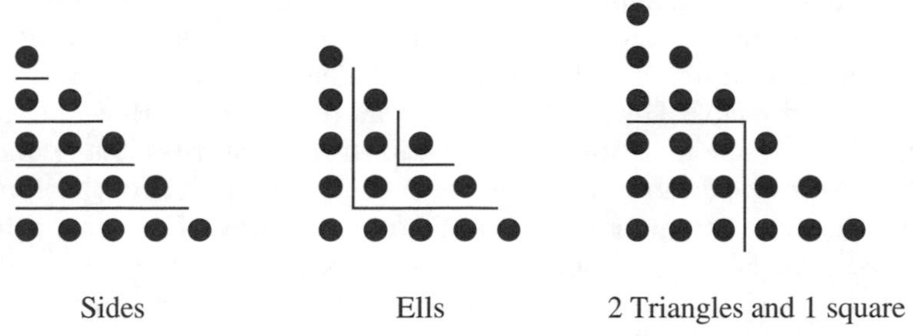

FIGURE 30 Multiple representations of a triangle.

6.3 Operations and Representations of Diagrams

The choice of representation that DIAMOND uses is important. Most of the proofs that DIAMOND proves require some kind of recursive decomposition of a diagram. If the appropriate representation of a diagram is available, then such decomposition is possible. Clearly, the more representations of a diagram are available, the more operations are possible on this diagram.

Given the multiple representations of diagrams defined in the previous section, we now list some of the operations that DIAMOND provides on particular diagrams. One of the features of DIAMOND is that it automatically restricts the operations offered to the user for each type of diagram. In this way, the user cannot carry out a nonsensical operation (e.g., to split a triangle into four squares). Here are the available operations for some diagrams:

Square: split a row, split a column, split an ell, cut diagonally, split an outer frame, split an inner dot, split into four squares.

Rectangle: rotate 90 degrees, split a row, split a column, split a square, cut diagonally.

Triangle: split a side, split an ell, split into two triangles and a square.

Ell: split row, split diagonal ends.

Thick Frame: split a frame, split into rectangles.

Notice, that we consider rotation as an explicit operation, hence we have distinct operations for splitting rows and columns. By the same principle, we should also have operations which, for example, split ells from a square in the other three directions. Indeed, these operations could be implemented in DIAMOND. However, for the theorems that we considered, these different orientations are not relevant, i.e., the proofs are independent of the orientation of ells. The justification for our choice of representations of diagrams and operations on them is heuristic. The particular sets of available diagrams and operations on them were selected by the analysis of examples of diagrammatic proofs, some of which are given in Chapter 3 and Appendix A. These sets can be extended which would enable Diamond to prove more theorems. However, the currently available diagrams and operations in Diamond already allow the user to prove a significant range and depth (see §9.2.1) of theorems.

A particular representation of a diagram is a way of viewing a diagram before enabling an operation to be carried out on it. For instance, if a square is viewed as a sequence of columns, then the operation that can be carried out on it is the recursive decomposition into a rectangle and a column. If a square is viewed as a nested sequence of ells, then the operation that is possible on it is the recursive decomposition into a smaller square and an ell.

New complex operations may emerge if these few representations that were presented in this chapter are combined in various ways. For instance, let a square be represented as four smaller squares, and we use any other type of representation for each of the four squares. This creates a new representation of a square, and as a consequence, allows a new complex recursive operation on a square. Amongst the available representations of diagrams the *recursive* representations in particular, give scope for many new recursive decompositions of diagrams.

Clearly, depending on a theorem and its proof, different operations are required. Consequently, diagrams need to be transformed into an appropriate representation. Sometimes, diagram representation needs to be transformed midway through the proof in order for the user to be able to use a particular operation. These transformations of diagram representations will be discussed next.

6.3.1 Transformation of Representations

Diamond has a notion of a default representation of diagrams. This representation is used when a diagram is first chosen. It is typically a matrix or a sequence of sides representation. As different operations are used, Diamond transforms the diagram into an appropriate representation.

Transformation between different representations are readily achieved and are invisible to the user. The idea is that the transformation of a diagram takes place behind the scenes, as it were, as the user chooses the operation. So for instance, say that the user wants to decompose a square into ells, then immediately a square is transformed into a representation of a nested sequence of ells. Using an appropriate representation enables easy handling of the operations. Figure 31 shows some of the transformations in the case of a square.

Sometimes, the transformation between two representations is not

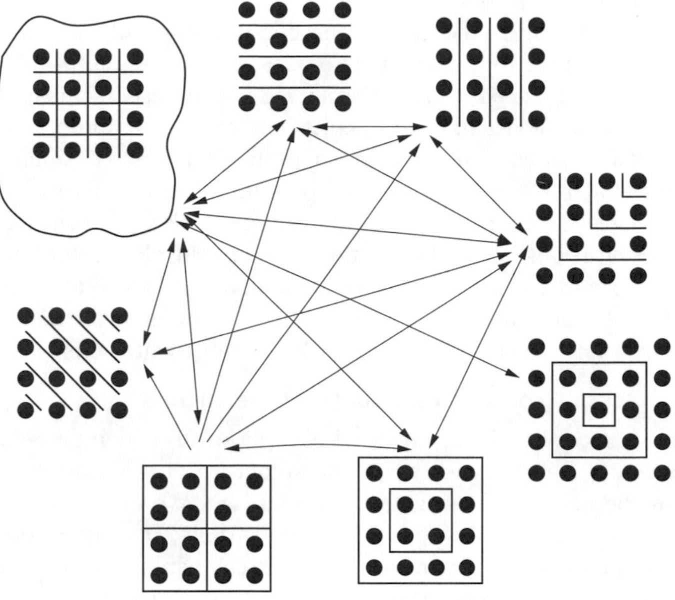

FIGURE 31 Transformation of representations of a square.

possible, hence the graph in Figure 31 is not fully connected. For instance, a square of odd magnitude cannot be transformed into the "four squares" representation of a square. On the other hand, if a representation is not available, then the operation on a diagram is not possible, unless the same operation can be composed of other operations. For example, consider the example-proofs for the theorem about the *sum of odd naturals* in Figure 24 and Figure 25 in the previous chapter. If a nested ell sequence representation of a square was not available, then the user could not split a square into an ell and a smaller square. However, the user could first split a row from a square (thus the square would be transformed into a row representation), which results in a rectangle and a row. Then the user could split a column from the resulting rectangle (thus a rectangle would be transformed into a column representation). This leaves the user with a square, where the two operations can be repeated. It is easy to see that if none of the three representations of a square and a rectangle were available, then the solution to the problem could not be found. It is obvious now how Pólya's advice about the importance of problem representation is used. A careful choice of a representation of the problem (i.e., diagrams) must be made in order to enable one to find a solution for it.

Notice that all of the operations could potentially be carried out using one diagram representation (e.g., a matrix representation). This would remove the need for transformations of representations. However, the operations on a single representation are much more complex in comparison to operations defined on multiple representations. Indeed, we argue that the operations should be simple, readily carried out, and should reflect the simplicity of human manipulation of diagrams. Multiple representations offer this advantage due to the close interdependence between a representation and an operation which is available and pertinent to this representation. This advantage outweighs the disadvantage of the need for transformations in multiple representations of diagrams.

6.3.2 Destructor v. Constructor Operations

We can see from the definitions that all operations decompose diagrams. Consequently, this is reflected in the structure of diagrammatic proofs that can be generated using these operations. We distinguish between destructor operations and constructor operations. Destructor operations decompose diagrams in various ways (i.e., they split diagrams into new diagrams), whereas constructor operations compose diagrams in various ways (i.e., they join diagrams into new diagrams). Hence, diagrammatic proofs can be of destructor and constructor type.[42] Since no differences appear in the usability of each type of operation in diagrammatic proofs, we choose to use only destructor operations in Diamond.

6.4 Operations as Tactics

Operations may be combined recursively into more complex ones. These combinations of operations are referred to as tactics. An atomic operation (such as splitting an ell from a square) is the simplest tactic. However, recursively applying this operation until the diagram is exhausted results in a more complex recursive tactic. Thus, tactics can use other tactics. To date the composition function for all of the composite operations is of the form "apply operations x, **then** apply y", where y is a recursive call in the composition function. There is scope to allow more complex tactics (e.g., consisting of conditional statements).

Figure 32 gives an example of how different tactics can be constructed from the example-proof for the theorem about the *sum of odd naturals*. Tactic 3 in Figure 32 consists of one atomic operation only. Tactic 2 uses Tactic 3 recursively to prove that an ell consists of an odd number of

[42] See Chapter 7 for a description of schematic proofs. Destructor schematic proofs are represented by the step case part first, followed by the base case part. Constructor schematic proofs would be represented in an opposite way.

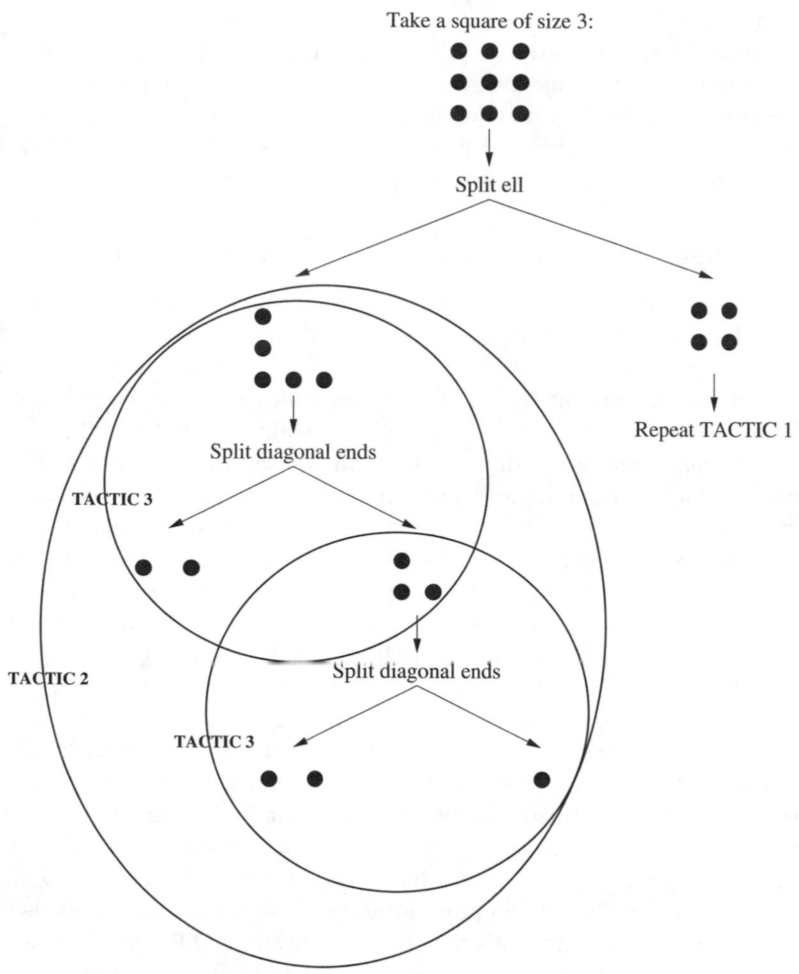

FIGURE 32 Operations as tactics.

dots. Tactic 1 recursively repeats the splitting of ell operation and Tactic 2 until the diagram is exhausted.

Notice that Figure 32 shows an example-proof rather than a general proof. It also represents how traces of example-proofs are stored in DIAMOND, so that a general proof can be constructed from them. A trace, i.e., a sequence of operations used in an example-proof, is recorded in DIAMOND using a tree structure. A linear sequence is mapped to a tree structure using a parameter which stores the position of the diagram in a sequence together with the operation which is applied to it. For instance, the operation "split ell" in Figure 32 has a parameter [ ] associated with it to indicate that it is applied to the initial diagram. The parameter for the first application of "split diagonal ends" is [1], and for its second application the parameter is [1, 2]. Construction of general proofs will be discussed in detail in Chapter 7.

6.5 Diagram Representation and Induction Schema

In §5.3.1 we showed how arithmetic expressions can be mapped into a diagrammatic representation. We also stated that a choice of a diagram and an operation on it fixes the choice of a mathematical induction schema. In this section we explain the relation between the choice of a diagram representation and an induction schema in more detail.

Choosing a representation of a diagram allows the use of a particular operation in the proof. Most diagrammatic proofs recursively decompose a diagram in some way. Carrying out each decomposition corresponds to doing a step case of a recursive decomposition. In a symbolic proof this process is analogous to the step case of mathematical induction. Therefore, choosing a representation of a diagram in a diagrammatic proof is analogous to choosing an induction rule in a symbolic proof.

The analysis of recursive definitions and their structures which suggests mathematical induction schemata and induction variables in the automation of inductive proofs dates back to Boyer and Moore (1979). Recursion analysis tries to find a suitable mathematical induction schema and universally quantified variables for induction. Sometimes there might be several induction schemata, or several variables over which to induct. If a wrong schema or variable is chosen, then the proof attempt is doomed to fail. In our diagrammatic proofs we define diagrams using different recursive definitions for the application of different geometric operations. We examine now examples of how various recursive representations of a square (and thus the operations on it) correspond to the choice of mathematical induction schemata in an analogous symbolic proof. We use mappings as described in §5.3.1 to make the correspondence between a diagrammatic and a symbolic proof explicit.

A square, when viewed as a collection of ells allows an lcut operation. If a square of magnitude n represents an arithmetic expression n^2 and an ell of magnitude n represents an arithmetic expression $2n-1$, then an lcut operation corresponds to the following rewrite rule: $s(n)^2 \Rightarrow n^2 + (2(s(n)) - 1)$ (we use a constructor s to define a successor function – see Glossary). Table 3 gives other possibilities according to the representation of a square, and consequently the operation on it.

TABLE 3 A square and the operations on it (n is a particular value).

	Operations and Diagrams
(1)	square of magnitude $\underline{s}(n)$ $\overset{\text{lcut}}{\Longrightarrow}$ square of magnitude n and ell of magnitude $s(n)$
(2)	square of magnitude $\underline{s(s(n))}$ $\overset{\text{split\_frame}}{\Longrightarrow}$ square of magnitude n and frame of magnitude $s(s(n))$
(3)	rectangle of magnitude $s(n) \times \underline{s}(n)$ $\overset{\text{split\_row}}{\Longrightarrow}$ rectangle of magnitude $s(n) \times n$ and row of magnitude $s(n)$
(4)	rectangle of magnitude $\underline{s}(n) \times s(n)$ $\overset{\text{split\_col}}{\Longrightarrow}$ rectangle $n \times s(n)$ and column of magnitude $s(n)$

The underlined parts of the expressions in Table 3 are the recursion constructors in the rules. They indicate the recursive argument which is used in the analysis to identify the induction schema. The same recursive constructors will be identified in the symbolic rewrite rules which correspond to the diagrammatic operations. To choose the rewrite rule which corresponds to a diagrammatic operation we need to select an appropriate mapping between the two. Let us use the following mappings for the arithmetic expressions:

- a square of magnitude n for n^2,
- a rectangle of length n and height m for $n \times m$,
- an ell of magnitude n for $2n - 1$,
- a row of magnitude n for n,
- a column of magnitude n for n,
- a frame of magnitude n for $4(n - 1)$.

Note that since DIAMOND only uses concrete diagrams, these mappings in a diagrammatic proof are given for particular values of n. We use a variable n to represent every instance in order to demonstrate the correspondence. The choices of diagram representation, and consequently the operations, given the mappings from diagrams to arithmetic expressions, correspond in a symbolic proof to the rewrite rules (again, note the underlined recursive constructors) presented in Table 4.

TABLE 4 Correspondence between diagrammatic and symbolic rules.

Diagrammatic Operation		Symbolic Rewrite Rule
(1)	lcut	$\underline{s}(n)^2 \Rightarrow n^2 + (2(s(n)) - 1)$
(2)	split_frame	$\underline{s(s}(n))^2 \Rightarrow n^2 + 4(s(s(n)) - 1)$
(3)	split_row	$s(n) \times \underline{s}(m) \Rightarrow s(n) \times m + s(n)$
(4)	split_col	$\underline{s}(n) \times s(m) \Rightarrow n \times s(m) + s(m)$

We may need a rule that says that for some n, a square of magnitude n is equal to a rectangle of magnitude $n \times n$ to be able to use the latter two rules. In the symbolic rewrite rules in Table 4 all the variables are implicitly universally quantified.

Consider again Table 3 and Table 4. Note how different recursive definitions of operations in Table 3 have different recursion constructions (they are underlined), which occur in the recursive argument positions. The same is the case in the analogous symbolic rewrite rules in Table 4 (they are also underlined). Operations (rewrite rules) (1), (3) and (4) have a one step recursive structure. Operation (rewrite rule) (2) has a two step recursive structure. Each of these recursion structures (schemata) corresponds to a mathematical induction schema. The one step induction schema is:

$$\frac{P(0) \qquad P(n) \vdash P(s(n))}{\forall n.\ P(n)}$$

The two step induction schema is:

$$\frac{P(0), P(s(0)) \qquad P(n) \vdash P(s(s(n)))}{\forall n.\ P(n)}$$

Choosing the representation of a square which allows (1), (3) or (4) fixes therefore, the choice of an induction schema to a one-step induction in the symbolic proof. Choosing the representation of a square which allows (2) fixes the choice to a two step induction schema. However, notice that choosing the representation of a square which allows (3) fixes the induction variable to be m, whereas choosing the representation to allow (4) fixes the induction variable to be n. In the diagrammatic proof the choice of possible operations, and thus representations of a diagram, is dependent on how we map the arithmetic expressions into a diagrammatic representation.

In summary, these examples demonstrate that the choice of a representation of a diagram and operations on them in a diagrammatic proof is analogous to fixing the choice of an induction schema and an induction variable in a symbolic proof.

On the other hand, it is interesting to notice that choosing an induction schema does not necessarily fix the choice of the representation that can be used for diagrams, and correspondingly the choice of possible operations. Rather, it restricts the set of possible diagram representations. For instance, were we to choose a two-step induction schema in the proof of a theorem, then the only representation that we could use for a square would be the one that allows operation (2) (i.e., split_frame in Table 4). Choosing a one-step induction schema restricts, but not uniquely determines our choice of representation. However, in some cases the choice of an induction variable may fix the choice of a diagram representation. For instance, if we choose m as an induction variable, then this fixes the representation of a square to be the one that allows operation (3).

6.6 Summary

Geometric operations form the inference steps of DIAMOND's proof. Their availability is closely related to the representation of a diagram. In particular, if no appropriate representation of a diagram is available, then the operation on a diagram might not be possible. Defining multiple representations of diagrams enabled us to identify all possible operations on each type of diagram.

Different operations are required in different proofs. Thus, the representation of a diagram sometimes needs changing midway through the proof. This transformation is readily achieved in DIAMOND and is invisible to the user. Using multiple representations and their transformations in diagrams sheds light on Pólya's advice on the importance of problem representation: while in symbolic reasoning there is usually only one type of representation available to the system, in diagrammatic reasoning the same diagrams can be represented using several representations. DIAMOND can readily choose the one that is most suitable for the problem at hand.

It is interesting to relate diagrammatic and symbolic proofs. It turns out that the choice of operations in the diagrammatic proof (and therefore the particular diagram representations) is analogous to the choice of a mathematical induction schema in a symbolic proof. The recursion analysis of definitions of diagrams identifies the recursion schema. Analogously, the recursion analysis of the definitions of rewrite rules for a symbolic proof identifies the induction schema to be used in an inductive symbolic proof.

A sequence of operations used in Diamond's example-proof is a tactic. Several such concrete tactics are abstracted into a general schematic proof. How an abstraction is carried out will be discussed in the next chapter.

7

The Construction of Schematic Proofs

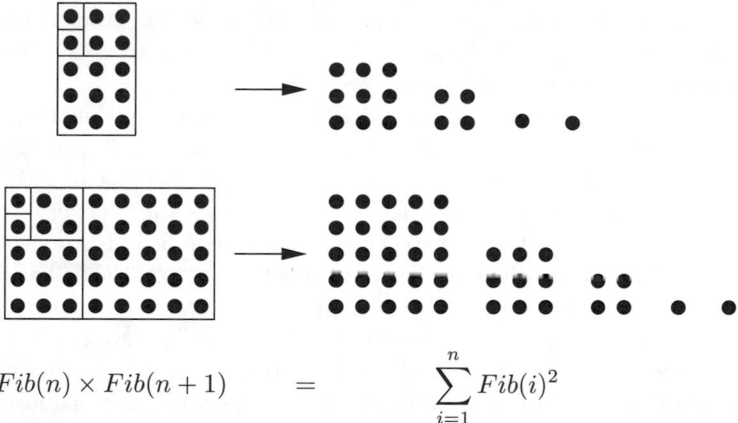

$$Fib(n) \times Fib(n+1) \quad = \quad \sum_{i=1}^{n} Fib(i)^2$$

— ALFRED BROUSSEAU
adapted from NELSEN's *Proofs Without Words*

The notion of a diagrammatic proof presented in this book is captured in a schematic proof. Schematic proofs are unary functions, i.e., programs, which output a proof of a proposition $P(n)$ for each value of their input n. In other words, they are a way of capturing a family of proofs. In this chapter we present how general schematic proofs are automatically constructed from example-proofs, and how they are formalized in DIAMOND.

We start in §7.1 by explaining what we mean by abstraction in the context of learning from examples. Schematic proofs are abstracted from example-proofs and this is discussed next in §7.2. In DIAMOND, schematic proofs have a particular formalization which is presented in §7.3. There are many existing abstraction techniques, some of these are more suitable in our domain than others – we list them in §7.4 with a particular focus on the treatment of the dependency between the number of steps and the instance under consideration. We are now in a position to define in §7.5 our abstraction mechanism for all linear dependency functions, which is further refined in §7.6. Sometimes, proof have case splits in them – we consider these next in §7.7. Finally, in §7.8 a possible mechanism for abstracting a general proof from *one* example-proof is discussed.

7.1 Context for Abstraction

In Chapter 4 schematic proofs were introduced. A *symbolic* schematic proof applies rewrite rules to construct a proof. The number of applications of the rewrite rules is dependent upon a parameter n. We argued that similarly to symbolic proofs, schematic proofs can be used for diagrammatic proofs. In a *diagrammatic* schematic proof, it is the geometric operations on a diagram which are used in the same way as the rewrite rules are in the symbolic schematic proof.

A schematic proof is constructed from examples of proofs of corresponding instantiations of the premises. As mentioned in §5.4, example-proofs are constructed interactively with the user. We let the user explore instances of proofs and we automate the construction of a general proof from these instances. This construction process is referred to as an abstraction of a general schematic proof from examples of proofs. The next two sections present how examples of proofs are stored in proof traces and how schematic proofs are formalized. Then, in §7.4 we analyze the some applicable abstraction methods with respect to the requirements for abstraction in DIAMOND. This analysis will enable us to choose the technique most suitable for our purposes.

7.2 Example-Proof Traces

A schematic proof of a theorem is constructed from a few example-proofs.[43] The construction of example-proofs was presented in §5.4. DIAMOND expects the example-proofs to be formulated in a particular way in order for it to be able to abstract from them. The aim is to recognize automatically the recursive structure of the proof from a linear sequence of applications of operations, so that the structure common to the example-proofs for n and $n+1$ can be recognized and abstracted into a general schematic proof. Notice that the example-proofs do not

[43]In DIAMOND we use two example-proofs, which is enough to be able to extract linear dependencies between the number of applications of geometric operations in the example-proofs (see §7.5). In §7.8 we discuss a possibility of abstracting from only one example-proof.

need to be given for adjacent values of n. This will be discussed further in §7.6.

Traces of example-proofs are recorded as sequences of applications of operations. For instance, take the two example-proofs given in Figure 24 and Figure 25 in Chapter 5. They are example-proofs for the theorem about the *sum of odd naturals* for the values of $n = 4$ and $n = 3$. The example-proof traces for $n = 4$ and for $n = 3$ consists of the following operations given in Table 5.

TABLE 5 Example-proof traces for $n = 4$ and $n = 3$ for *sum of odd naturals*.

Value of $n = 4$	
Operation	No. of applications
lcut	1
split_ends	3 (i.e., 4-1)
lcut	1
split_ends	2 (i.e., 3-1)
lcut	1
split_ends	1 (i.e., 2-1)
lcut	1
split_ends	0 (i.e., 1-1)
Value of $n = 3$	
Operation	No. of applications
lcut	1
split_ends	2 (i.e., 3-1)
lcut	1
split_ends	1 (i.e., 2-1)
lcut	1
split_ends	0 (i.e., 1-1)

The sectioning of the tables indicates the structure common to the two example-proofs. This structure needs to be automatically detected by DIAMOND, and is reformulated into the following representation:

$$\begin{array}{lrcr} n = 4: & \mathsf{proof}(4) & = & \mathcal{A}(4)\mathcal{A}(3)\mathcal{A}(2)\mathcal{A}(1)\mathcal{B}(0) \\ n = 3: & \mathsf{proof}(3) & = & \mathcal{A}(3)\mathcal{A}(2)\mathcal{A}(1)\mathcal{B}(0) \end{array}$$

where $\mathcal{A}(i)$ is the step case of the proof and consists of some sequence of operations (in the example above these are lcut and split_ends) and $\mathcal{B}$ is a base case which also consists of some sequence of applications of operations or is empty (as in the example above). The index i denotes the value of n for each particular step case. The sequence of operations and the number of applications of operations in the step case is dependent on the case of the proof, i.e., the value of n.

In a more general case of example-proofs for n and $n+1$ the representation can be reformulated into the following:

$$\begin{array}{lcr} \mathsf{proof}(n) & = & \mathcal{A}(n), \mathcal{A}(n-1), \mathcal{A}(n-2), \ldots, \mathcal{A}(1), \mathcal{B} \\ \mathsf{proof}(n+1) & = & \mathcal{A}(n+1), \mathcal{A}(n), \mathcal{A}(n-1), \mathcal{A}(n-2), \ldots, \mathcal{A}(1), \mathcal{B} \end{array}$$

It is possible that a theorem does not have a proof for all consecutive values of n, but rather for all odd or even, or any other subset of values of n. Thus, in a case of two example-proofs, the representation of a schematic proof can be reformulated for any natural number c and n into the following (r is a remainder in $n \ mod \ c$):

$$\begin{array}{lcr} \mathsf{proof}(n) & = & \mathcal{A}(n), \mathcal{A}(n-c), \mathcal{A}(n-2c), \ldots, \mathcal{A}(c+r), \mathcal{B} \\ \mathsf{proof}(n+c) & = & \mathcal{A}(n+c), \mathcal{A}(n), \mathcal{A}(n-c), \mathcal{A}(n-2c), \ldots, \mathcal{A}(c+r), \mathcal{B} \end{array}$$

We can now formalize the recursive function proof.

7.3 Formalization of Schematic Proofs

We are interested in inductive diagrammatic proofs. More precisely, we consider proofs for $n+c$ which can be reduced to proofs for n (or conversely, such proofs for n which can be extended to proofs for $n+c$ by adding to them some additional sequence of operations). It is precisely this difference between the $\mathsf{proof}(n+c)$ and $\mathsf{proof}(n)$, i.e., the additional sequence of operations in $\mathsf{proof}(n+c)$ with respect to $\mathsf{proof}(n)$ that we call the step case of the abstracted proof.

Sometimes proofs are not uniform for all values of n. A theorem could have different schematic proofs for say, even and odd natural numbers. Such a proof clearly contains a case split: there is one schematic proof for odd naturals and another schematic proof for even naturals. The abstraction mechanism that will be described in the next section has to detect when a proof has a case split or not. If a schematic proof is the same for all values of n, i.e., there is only one case of the proof, so $c = 1$, we seek the following recursive reformulation of a schematic proof.

$$\mathsf{proof}(n+1) = \mathcal{A}(n+1),\ \mathsf{proof}(n) \tag{7.4}$$

$$\mathsf{proof}(0) = \mathcal{B} \tag{7.5}$$

Note that $\mathsf{proof}(0)$ is often an empty list of operations, because often no diagram is defined for $n = 0$, i.e., a diagram which consists of no dots.

Proofs that have the same structure for all n are called 1-homogeneous proofs. Proofs can be c-homogeneous; then there are c cases of the proof. We say that if all instances of the proof (for instances of numbers that "equal modulo c") have the same structure and can be abstracted, then the proof is c-homogeneous. If there are c cases, then there are c different abstracted proofs, one for each case. We seek the smallest complete recursive definition of a proof, i.e., c potentially different schematic proofs, if there are c cases. The following theorem and corollary will help us define what we mean by the smallest complete proof:

Theorem 1 *If a proof is c-homogeneous, then it is also (kc)-homogeneous for every natural number $k > 0$.*

The immediate consequence of Theorem 1 is:

Corollary 2 *If a proof is* not *c-homogeneous, then it is also* not *f-homogeneous for every factor f of c.*

In a c-homogeneous proof we will denote by $\mathcal{B}_r$ a base case for a branch of numbers which give remainder r when divided by c. $\mathcal{B}_r$ is actually a proof for the smallest natural number that gives remainder r when divided by c.

A schematic proof is defined to be the smallest complete proof if there is no other f-homogeneous proof obtainable from a c-homogeneous proof for any factor f of c, and all f schematic proofs for f cases are defined.

The general representation of a *destructor*[44] proof is formalized as follows – let:

- $n = kc + r$
- where $c =$ *number of cases* and $r < c$
- and $i \geq 1$.

Then the recursive definition of a general proof is:

$$\begin{aligned} \mathsf{proof}(ic+r) &= \mathcal{A}_r(ic+r),\ \mathsf{proof}((i-1)c+r) \\ \mathsf{proof}(r) &= \mathcal{B}_r \end{aligned}$$

where $\mathcal{A}_r$ is a step case and $\mathcal{B}_r$ is a base case for a class of proofs where $n \equiv r(mod\, c)$. "," denotes a concatenation of operations in $\mathcal{A}_r$ and proof. The formalization of abstracted proof for *constructor* proofs is symmetric to the one given above.

Note that more complex proof structures are possible, e.g.,

$$\begin{aligned} \mathsf{proof}(n+1) &= \mathcal{A}(n+1),\ \mathsf{proof}(n),\ \mathcal{A}'(n+1) \\ \mathsf{proof}(0) &= \mathcal{B} \end{aligned}$$

However, proofs with such complex proof structures are very rare, and we do not treat them in this book.

[44]The notion of destructor and constructor proofs has been introduced in §6.3.2.

7.4 Abstraction Techniques

The term "abstraction" is used in this book to refer to the process of inferring general arguments from specific ones. Abstraction summarizes a set of data in a way, so that the new representation can predict new instances of the data set. In particular, we refer to learning from examples, i.e., using a set of examples to find a model that fits all the instances of our set. Sometimes mathematical models are used for this. Using the model, we can infer new instances. We are in particular interested in abstracting a general proof from instances of a proof. This is not a new problem — many abstraction techniques have been around for a few decades (Plotkin 1969, Winston 1975, Mitchell 1978, Michalski 1983).

One of the first and most influential algorithms for abstraction was introduced by Plotkin and is known as least general generalization (LGG) (Plotkin 1969, 1971). Here, we list some existing abstraction algorithms and refer the reader to the cited literature for more details.

- Abstraction mechanism by Biermann (1972): an algorithm which learns from examples of execution traces of a program and thus synthesizes the program.
- Bauer (1979) extended Biermann's program synthesis algorithm to make it more powerful and general. At the same time he used ideas from LGG to abstract a program from its execution traces.
- Abstraction algorithm by Anderson and Kline (1979): it uses examples as well as counterexamples to construct an abstraction.
- Version spaces by Mitchell (1982): there are two sets of descriptions – the set of general descriptions is modified by general-to-specific method; the set of specific descriptions is modified by specific-to-general method. The algorithm finds the abstraction when both sets are the same.
- Decision trees by Quinlan (1986): examples are represented as lists of attribute-value pairs.
- Inductive logic programming (ILP) by Muggleton (1991): an alternative approach to abstraction. It combines techniques of logic programming and inductive learning. ILP is used to synthesize new knowledge.
- Abstraction mechanism by Baker (1993): abstracts general proofs from example-proofs in the domain of arithmetic proofs. One of the most important features of Baker's abstraction is the ability to construct a general function which by instantiation generates the examples it was constructed from.

There are features of Biermann's and Baker's algorithm that we can use in DIAMOND's abstraction mechanism. In particular, Biermann's algorithm detects the recursive structure of the example. This mechanism needs to be extended so that the system automatically detects the recursive structure. Baker's abstraction of dependency functions is also a feature which we can use in our abstraction algorithm. However it needs to be extended to detect complex dependency functions rather than fixed ones defined by Baker. Our abstraction mechanism is described in detail next.

7.5 Abstracting for All Linear Functions

As mentioned above, we aim to recognize a recursive structure of the given example-proofs. More precisely, we want to construct the step case $\mathcal{A}$ and the base case $\mathcal{B}$ of the proof and then abstract them for all n. The general methodology employed for doing this can be demonstrated as in Figure 33, where Y is the whole of an example-proof for a particular

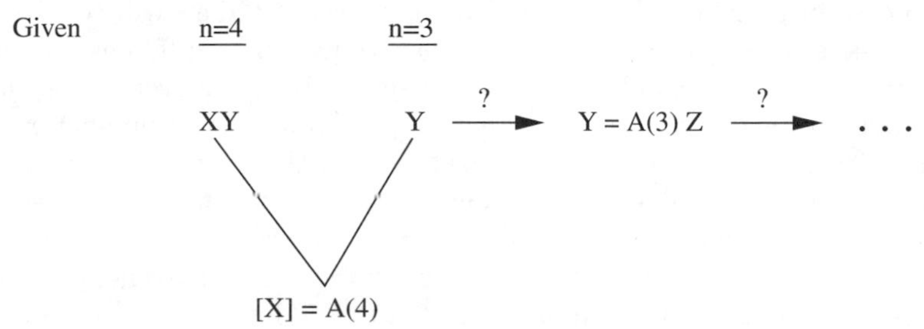

FIGURE 33 Methodology for abstraction mechanism.

n (in this case for $n = 3$) and X is the difference between example for $n+1$ and n (in this case, the difference between example-proofs for $n = 4$ and $n = 3$). X and Y consist of sequences of applications of geometric operations. The difference X is a step case $\mathcal{A}$ of the schematic proof for a particular value of n (in this case $n = 4$).

The first step of the abstraction algorithm is to extract the difference between the two example-proofs for n_1 and n_2 ($n_1 > n_2$), where $c = n_1 - n_2$, in the hope that this, when abstracted, will be the step case $\mathcal{A}$ of the proof. Note that if there is more than one case of the proof, say there are c cases, then n_1 and n_2 need to be given for the same case of the proof. There will also be c different step cases $\mathcal{A}$, one for each case. The construction of $\mathcal{A}$ is done by commutative and associative matching which detects and returns the difference between the two example-proofs.[45] Now we have a concrete step case of the proof. This difference consists of a few, say m, operations op_k each applied x_{k,n_1} times for some natural number k, where $0 < k \leq m$.

To make a step case general, we need to find the dependency function between every x_{k,n_1} and n_1. This demands identifying a function of n_1, which gives a specific x_{k,n_1}, i.e., $f_k(n_1) = x_{k,n_1}$ for some k and n_1. DIAMOND assumes that the dependency is linear: of the form $an + b$. Examples show that this is a heuristically adequate choice. Thus, let us write for each op_k a linear equation $an_1 + b = x_{k,n_1}$, where n_1 and x_{k,n_1} are known. Note that DIAMOND cannot cope with, for instance, exponential, logarithmic or polynomial functions.[46]

The subsequent stage of the abstraction is to construct the next step case from the rest of the example-proof for the corresponding new n (i.e., n_2). If successful, continue constructing step cases for the corresponding n's from the rest of the proof until only the base case is left.

Since we are dealing with inductive proofs, it is expected that every step case of a proof will have the same structure,[47] i.e., will consist of the same sequence of application of operations, but a different number of times. Thus, we could in the same way as above for every operation op_k write a linear equation $an_2 + b = x_{k,n_2}$. However, the number x_{k,n_2} of applications of a particular operation op_k in the next step case is not known. A possible value of x_{k,n_2} is acquired by counting the number x' of times that every operation op_k of the initial step case occurs in the rest of the proof. The actual value of the number of occurrences of each operation could be any number from 0 to x'. Thus, we branch for all such values and so we have:

$$\begin{array}{rcl} an_1 + b & = & x_{k,n_1} \\ an_2 + b & = & x_{k,n_2} \end{array}$$

where n_1, n_2, x_{k,n_1} and x_{k,n_2} are known, so the equations can be solved for a and b, and x_{k,n_2} takes values from 0 to x'. This results in several possible potential abstractions of the step case, where branching involves solving the following equations for each operation of the step case:

[45] Using commutative and associative matching reduces sensitivity to the order of proof steps.

[46] However, DIAMOND could be extended to deal not only with linear, but also more complex dependency functions such as exponential or polynomial ones.

[47] Recall that if there is a case split in the proof, then the step cases of the same case of the proof will have the same structure. However, step cases of different cases of the proof might differ.

$$an_1 + b = x_{k,n_1}$$

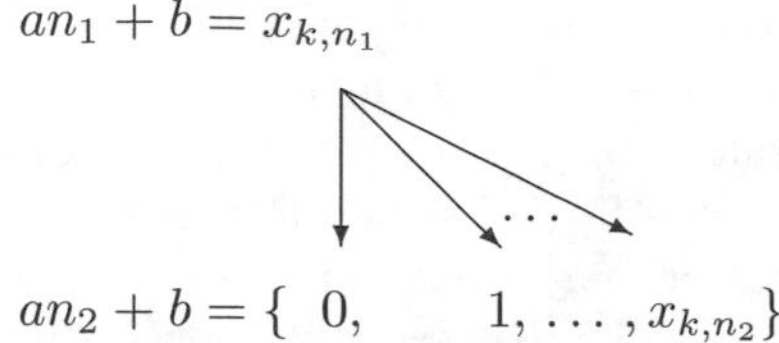

$$an_2 + b = \{\ 0, \quad 1, \ldots, x_{k,n_2}\}$$

The aim is to eliminate those that are impossible. After checking if step cases for all n down to the base case are structurally consistent (i.e., the number of applications of geometric operations is as expected by instantiating the chosen dependency function) one hopes to be left with at least one possible abstraction of the example-proofs. The step case is rejected when the sequence of operations in the subsequent step cases is impossible, i.e., the functions were chosen incorrectly. This normally occurs when the dependency function gives a negative number of applications of a particular operation, when the calculated sequence is not identical to the rest of the example-proof, or when there is no integer solution to our equations. Usually, there will be only one possible abstraction of the two given example-proofs.

The example-proof for the *sum of odd naturals* is abstracted to form the following step case and base case:

$$\begin{aligned} \mathcal{A}(n) &= [(\mathsf{lcut}, 1), (\mathsf{split\_ends}, n-1)] \\ \mathcal{B} &= [\,] \end{aligned}$$

where the function in parentheses indicates the number of times that the operations are applied for each particular n. Thus, the following is the schematic proof for the theorem about the *sum of odd naturals*:

$$\begin{aligned} \mathsf{proof}(n+1) &= [(\mathsf{lcut}, 1), (\mathsf{split\_ends}, n)],\ \mathsf{proof}(n) \\ \mathsf{proof}(0) &= [\,] \end{aligned}$$

7.5.1 Example of Abstraction

Consider the example-proof traces for the theorem about the *sum of odd naturals* given in Table 5. We give here an example of how to abstract a schematic proof from the two example-proof traces.

The first step is to detect the difference between the two example-proofs. In our case this is $\mathcal{A}(4) = [(\mathsf{lcut}, 1), (\mathsf{split\_ends}, 3)]$. Thus we have $n_1 = 4$, $op_1 = \mathsf{lcut}$, $x_{1,4} = 1$, and $op_2 = \mathsf{split\_ends}$, $x_{2,4} = 3$.

Next, we need to find dependency functions between $n_1 = 4$ and $x_{1,4} = 1$, and $n_1 = 4$ and $x_{2,4} = 3$, i.e., we need to find functions f_1 and f_2 such that $f_1(4) = 1$ and $f_2(4) = 3$. We assume that the dependency function is linear: $n_1 a + b = x_{k,n_1}$. Thus we have: $f_1(4) = 4a + b = 1$ and $f_2(4) = 4a + b = 3$.

The subsequent stage is to construct the next step case from the two example-proofs. We seek the linear dependency function $n_2 a + b = x_{k,n_2}$ for each operation op_k. The value of n_2 is known ($n = 3$), but x_{k,n_2} can take any value from 0 to x'. Recall that x' is the number of times that the operation op_k occurs in the rest of the example-proof. So for op_1 = lcut, x' is 2. For op_2 = split_ends, x' is 3. Therefore the possible functions for op_1: $3a + b = 0$, $3a + b = 1$ and $3a + b = 2$. Figure 34 shows the system of two equations which need to be solved to find the

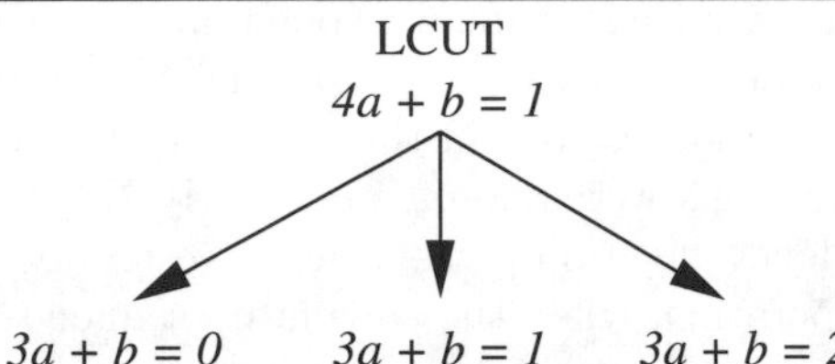

FIGURE 34 Branching of dependency function for lcut.

dependency function for op_1.

For op_2 the possible functions are $3a + b = 0$, $3a + b = 1$, $3a + b = 2$ and $3a + b = 3$. Figure 35 shows the system of two equations which need to be solved to find the dependency function for op_2.

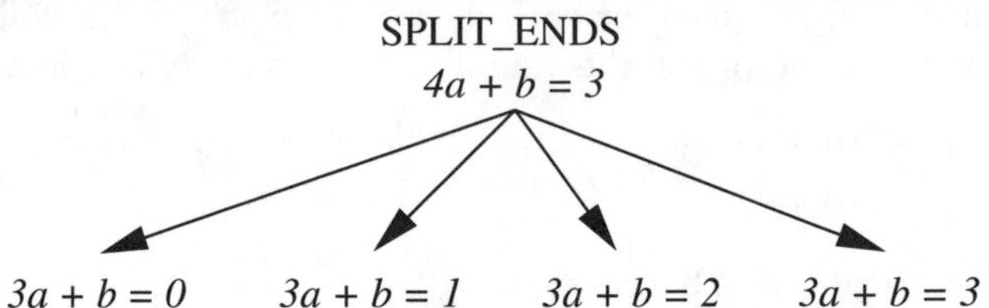

FIGURE 35 Branching of dependency function for split_ends.

Solving the system of two equations for op_1 to get the values for a and b, we get the following possible functions f_1:

- $f_1(n) = n - 3$
- $f_1(n) = 1$
- $f_1(n) = 5 - n$

For op_2 the get the following possibilities for f_2:

- $f_2(n) = 3n - 9$
- $f_2(n) = 2n - 5$
- $f_2(n) = n - 1$
- $f_2(n) = 3$

Instantiating these functions for any value of $n \leq 4$ and checking it with any of the two actual example-proofs eliminates impossible functions and identifies that $f_1(n) = 1$ is the dependency function for $op_1 = \mathsf{lcut}$, and $f_2(n) = n - 1$ is the dependency function for $op_2 = \mathsf{split\_ends}$.

7.6 Breaking c-Homogeneous to f-Homogeneous Proof

Consider again the two example-proofs for the *sum of odd naturals* (the example-proof consists of making n lcuts, and then showing that each ell consists of an odd number of dots). If the user supplies two example-proofs for values of n and $n + 1$, for some concrete n, then there is no problem, so DIAMOND will abstract normally and determine that the proof is 1-homogeneous. However, should the user supply proofs for n and $n + 2$ for some concrete n, then the first stage of abstraction determines that the step case consists of two lcuts. However, a complete recursive function for abstraction requires a step case to consist of one lcut only.

For instance, suppose a user supplies examples for some concrete n and $n+2$ where n is an even number in the proofs for the theorem about the *sum of odd naturals* and the abstraction mechanism described so far constructs the following schematic proof:

$$\begin{aligned} \mathsf{proof}(n+2) &= [(\mathsf{lcut}, 1), (\mathsf{split\_ends}, n+1), (\mathsf{lcut}, 1), (\mathsf{split\_ends}, n)], \\ & \quad \mathsf{proof}(n) \\ \mathsf{proof}(0) &= [\,] \end{aligned}$$

Some inspection of the schematic proof indicates that this, first, is not a complete definition (i.e., there is no definition of a schematic proof for odd natural numbers), and second, that the step case of this schematic proof can be further broken down into the following:

$$\begin{aligned} \mathcal{A}(n+2) &= [(\mathsf{lcut}, 1), (\mathsf{split\_ends}, n+1)] \\ \mathcal{A}(n+1) &= [(\mathsf{lcut}, 1), (\mathsf{split\_ends}, n)] \\ \vdots \quad & \vdots \quad \vdots \\ \mathcal{A}(2) &= [(\mathsf{lcut}, 1), (\mathsf{split\_ends}, 1)] \\ \mathcal{A}(1) &= [(\mathsf{lcut}, 1), (\mathsf{split\_ends}, 0)] \\ \mathcal{B} &= [\,] \end{aligned}$$

This can be recursively re-defined as:

$$\begin{aligned}\mathsf{proof}(n+1) &= [(\mathsf{lcut}, 1), (\mathsf{split\_ends}, n)],\ \mathsf{proof}(n)\\ \mathsf{proof}(0) &= [\,]\end{aligned}$$

which is what we expect. This recursive definition is now a complete, i.e., defined for all natural numbers, and the smallest schematic proof.

Diamond has a mechanism which detects whether a schematic proof can be further broken down, as in the example just given. It checks this by trying to split the step case into a further f structurally same sequences of operations, for all factors f of c in order to obtain an f-homogeneous proof. We give now a method for construction of an f-homogeneous proof from a c-homogeneous proof, where f is a factor of c. An example of using this method to break down the step case for *sum of odd naturals* follows in §7.6.1.

Let $\mathcal{A}(n)$ be the step case of the abstracted schematic proof, consisting of some sequence of operations. The number of applications for each operation is expressed as a dependency function on n. The algorithm consists of the following steps:

1. For each operation op_k count how many times it occurs in $\mathcal{A}(n)$. Therefore, we have $occ(op_k) = an + b$, where a and b are known.

2. For each factor f of c, assume that each operation op_k occurs $\alpha(n - lf) + \beta$ times for l ranging from 0 to m, where m is such that $mf < c$, more precisely, $(m+1)f = c$ and thus $m = \frac{c}{f} - 1$. Therefore we have:

$$\begin{aligned}occ_0(op_k) &= \alpha n + \beta\\ occ_f(op_k) &= \alpha(n - f) + \beta\\ &\vdots\\ occ_{lf}(op_k) &= \alpha(n - lf) + \beta\\ &\vdots\\ occ_{mf}(op_k) &= \alpha(n - mf) + \beta\end{aligned}$$

3. For each operation op_k, and for each factor f of c, add all of the above equations. After some simplification we get:

$$\begin{aligned}
&\sum_{m=0}^{\frac{c}{f}-1} \alpha(n-mf)+\beta \\
&= (\frac{c}{f}\alpha)n + ((-(0+f+2f+\cdots+(\frac{c}{f}-1)f))\alpha + (\frac{c}{f})\beta) \\
&= (\frac{c}{f}\alpha)n + ((-f(0+1+2+\cdots+\frac{c-f}{f}))\alpha + (\frac{c}{f})\beta) \\
&= (\frac{c}{f}\alpha)n + ((-f(\frac{\frac{c-f}{f}(\frac{c-f}{f}+1)}{2}))\alpha + (\frac{c}{f})\beta) \\
&= (\frac{c}{f}\alpha)n + ((-f((\frac{1}{2})(\frac{c-f}{f})(\frac{c}{f})))\alpha + (\frac{c}{f})\beta) \\
&= (\frac{c}{f}\alpha)n + ((-\frac{c(c-f)}{2f})\alpha + (\frac{c}{f})\beta)
\end{aligned}$$

where f and c are known.

4. For each operation op_k, solve the system of equations in 1.) and 3.) for α and β. Thus:

$$(\frac{c}{f}\alpha)n + ((-\frac{c(c-f)}{2f})\alpha + (\frac{c}{f})\beta) = an + b$$

where a, b, c and f are known. Thus, by equating the coefficients of n and 1 on both sides we get:

$$\alpha = a\frac{f}{c} \qquad \text{and} \qquad \beta = \frac{b + \frac{c(c-f)}{2f}}{\frac{c}{f}}$$

5. For these α and β, solve the equations of 2.), which results in the number of occurrences of each operation op_k for a particular factor f in a corresponding part of the divided step case $\mathcal{A}(n)$. Furthermore, for each divided part of the step case, the order of operations has to be preserved from the original step case $\mathcal{A}(n)$.

Using this algorithm, one can determine where to split the step case $\mathcal{A}(n)$ into f structurally identical parts.

If the method fails, then there is no such f-homogeneous further abstraction of the step case $\mathcal{A}(n)$. If the method succeeds, and DIAMOND finds a new abstraction of the step case, call this $\mathcal{A}'(n)$, then it also needs to find a new base case $\mathcal{B}'_{r'}$ if $r' \neq 0$, or $\mathcal{B}'_f$ if $r' = 0$, where the previous r for c was such that $n = kc + r$ and $r < c$, and the new r' is now such that $n = kf + r'$ and $r' < f$. The proofs of soundness and completeness (given the limitation of the algorithm, e.g., linear dependency function restriction) of this abstraction algorithm can be obtained by appealing to the construction of the algorithm.

7.6.1 Example of Abstracting an f-Homogeneous Proof

Consider again, the example of a schematic proof for which a step case $\mathcal{A}$ consists of the following operations:

$$\mathcal{A}(n) = [(\mathsf{lcut}, 1), (\mathsf{split\_ends}, n-1), (\mathsf{lcut}, 1), (\mathsf{split\_ends}, n-2)]$$

where the number in brackets indicates the number of applications of that particular operation. Assume also that DIAMOND determined after the first abstraction attempt that the proof is 2-homogeneous, i.e., it has two cases. We demonstrate now how the algorithm in the previous section splits the step case of the proof further so that the function proof becomes 1-homogeneous.

Recall from the previous section that we want the step case to be $\mathcal{A}(n) = [(\mathsf{lcut}, 1), (\mathsf{split\_ends}, n-1)]$ (while as expected in order to preserve the structure of the proof we have $\mathcal{A}(n-1) = [(\mathsf{lcut}, 1), (\mathsf{split\_ends}, n-2)]$).[48]

The algorithm splits the step case $\mathcal{A}$ into f parts, yet retains the same structure for all split parts in terms of dependency on n. Consider now how this algorithm works for the example just given, where $c = 2$ and $f = 1$.

1. $occ(\mathsf{lcut}) = 2$ where $a = 0$ and $b = 2$, and $occ(\mathsf{split\_ends}) = (n-1) + (n-2) = 2n - 3$ where $a = 2$ and $b = -3$.
2. We have:
   (a) $occ_0(\mathsf{lcut}) = \alpha_1 n + \beta_1$,
   (b) $occ_1(\mathsf{lcut}) = \alpha_1(n-1) + \beta_1$,
   (c) $occ_0(\mathsf{split\_ends}) = \alpha_2 n + \beta_2$,
   (d) $occ_1(\mathsf{split\_ends}) = \alpha_2(n-1) + \beta_2$.
3. Note that $m = \{0, 1\}$ so that $mf < c$ $(f = 1, c = 2)$. Then:

$$\mathsf{lcut} \longrightarrow \sum_{m=0}^{1} \alpha_1(n - mf) + \beta_1 = (2\alpha_1)n + ((-1)\alpha_1 + (2\beta_1))$$

$$\mathsf{split\_ends} \longrightarrow \sum_{m=0}^{1} \alpha_2(n - mf) + \beta_2 = (2\alpha_2)n + ((-1)\alpha_2 + (2\beta_2))$$

[48] Note that in §7.6 we defined the step case $\mathcal{A}$ for $n+2$, whereas here we define it for n where n is even. Essentially we are considering the preceding instantiation of the recursive call in the schematic proof. This is due to the mechanism being defined for $\mathcal{A}(n)$ rather than $\mathcal{A}(n+2)$. Renaming of variables could be used instead, e.g., $n+2$ can be renamed into m so the algorithm applies to $\mathcal{A}(m)$.

Thus:

$$\begin{aligned}(2\alpha_1)n + ((-1)\alpha_1 + (2\beta_1)) &= 2\\(2\alpha_2)n + ((-1)\alpha_2 + (2\beta_2)) &= 2n-3\end{aligned}$$

So $\alpha_1 = 0$, $\beta_1 = 1$, and $\alpha_2 = 1$, $\beta_2 = -1$.

4. Now we have:
   (a) $occ_0(\mathsf{lcut}) = 0n + 1 = 1$,
   (b) $occ_1(\mathsf{lcut}) = 0(n-1) + 1 = 1$,
   (c) $occ_0(\mathsf{split\_ends}) = 1n - 1 = n - 1$,
   (d) $occ_1(\mathsf{split\_ends}) = 1(n-1) - 1 = n - 2$.

   Therefore following the order of operations in the initial step case, we now have: $\mathcal{A}(n) = [(\mathsf{lcut}, 1), (\mathsf{split\_ends}, n-1)]$ (while as expected $\mathcal{A}(n-1) = [(\mathsf{lcut}, 1), (\mathsf{split\_ends}, n-2)]$). Hence the schematic proof can now be re-defined as:

$$\begin{aligned}\mathsf{proof}(n+1) &= [(\mathsf{lcut}, 1), (\mathsf{split\_ends}, n)],\ \mathsf{proof}(n)\\\mathsf{proof}(0) &= [\,]\end{aligned}$$

7.7 Proofs With Case Splits

A theorem could have structurally different schematic proofs for different classes of values n. Such a proof contains a case split. The abstraction mechanism described in §7.6 can deal with proofs that have uniform case splits, i.e., proofs that have different structure for:

- 2 cases: classes of numbers that are:
  divisible by 2 (even)
  giving rest=1 when divided by 2 (odd)
- 3 cases: classes of numbers that are:
  divisible by 3
  giving rest=1 when divided by 3
  giving rest=2 when divided by 3
- 4 cases: classes of numbers that are:
  divisible by 4
  giving rest=1 when divided by 4
  ⋮

so the proofs are said to be 2-homogeneous, 3-homogeneous, 4-homogeneous, and so on, respectively (where a 1-homogeneous proof is trivial with one case only). As shown in §7.6 DIAMOND can detect such linear case splits. However, the system cannot deal with case splits that are homogeneous for any other *non-linear* sequence of numbers (e.g., exponential, logarithmic, prime, etc.).

Suppose the user constructs two example-proofs for some particular values n and $n + c$. As described in §7.6 Diamond first abstracts the recursive function $\mathsf{proof}(n)$ with c as the difference in the value of n in subsequent recursive calls. Then it reduces this c-homogeneous proof into an f-homogeneous schematic proof, where f is a factor of c, if such a proof exists. If there are no possible reductions to f-homogeneous proof, then the proof is c-homogeneous and by Corollary 2 there is no f-homogeneous further abstraction of the proof. Furthermore, if a proof is c-homogeneous, then Diamond requests from the user to supply $2 \times (c - 1)$ additional example-proofs in order to be able to abstract them for the other branches of the case split, and make the recursive function proof which represents the schematic proof total. Note that for each branch of the case split, the pairs of additional example-proofs have to be a factor f of c, or a multiple of f apart.

Suppose now, that a theorem does contain a case split, i.e., it is c-homogeneous and $c \neq 1$, but the user supplies two example-proofs that are not for the same case of the proof (i.e., not for n and $n + kc$, for some particular values of n and any multiple of c, say kc). Clearly, Diamond cannot abstract these example-proofs to form a schematic proof, because no such schematic proof exists. When Diamond fails to abstract a schematic proof from the given examples, then there are several reasons to which it can draw the user's attention. For example, one of them is that the two example-proofs are given for different cases, so Diamond can suggest to the user to supply another example-proof for each case, in order for it to be possibly able to abstract.

7.8 Abstracting From One Example

Is it possible to abstract a general schematic proof from just one example-proof? The explanation-based generalization[49] provides a mechanism for doing just that (Mitchell et al. 1986, DeJong and Mooney 1986). It is a technique which enables a formulation of general concepts from one specific training example. It differs from other inductive abstraction techniques in that it ever only needs one example to abstract from. The basic idea of a system that uses explanation-based generalization is that the system constructs explanations of why an object satisfies a function definition. It employs a domain model. A domain model is used to construct the explanation of why the training example satisfies the function definition. Then, the training example is transformed using this explanation into the most general form (usually by replacing constants

[49]The term "generalization" in explanation-based generalization means abstraction in our terminology.

with variables). The problem with this type of abstraction technique is that a considerable domain knowledge needs to be available before any abstraction can take place. It is a deductive rather than an inductive learning method. In a diagrammatic reasoning system there is no such extensive domain knowledge available in advance. It also cannot be built into the system, because it does not exist prior to carrying out the examples. The entire principle is based on the fact that a diagrammatic proof is induced from a set of examples without any prior knowledge of what the proof should look like.

For instance, one of the requirements in DIAMOND is to abstract the dependency functions from the proof applications. Recall that the dependency function defines the dependency between the parameter n for which a schematic proof is given, and the number of applications of particular geometric operations. Consider, for instance, that the training example was given for $n = 2$ and the number of applications of a particular operation is 4. The dependency functions for these values could be: $f(n) = 4, f(n) = 2n, f(n) = n^2$. Which one is the right one? There is no piece of domain knowledge which could determine the preference of one function over another. Were we to provide another example where $n = 3$ and the number of applications of the same operation is 6, then the only choice from the ones given is $f(n) = 2n$. It seems, therefore, that explanation-based generalization is not enough to induce a general diagrammatic proof from examples. There is not enough domain knowledge to abstract, or if we define such knowledge (e.g., we pick the preference for the given functions randomly), we might abstract incorrectly.

It could be argued, however, that humans do see a pattern from just one trusted example. This is probably more true for simple examples where the recursive structure of the proof is transparent. We explore this feature of human "informal" reasoning by enabling a system to exploit the hint about the recursive structure of the proof given by the user. This allows the system to abstract a general schematic proof of a theorem from one example only.

Consider the example-proof for the theorem about the *sum of odd naturals* where $n = 4$, given in Figure 24 in Chapter 5 and the corresponding example-proof trace given in Table 5. The recursive structure of the proof is inherent in the recursive structure of a square. If we take a square of magnitude four and split and ell from it, and then split end dots from the ell to show that it consists of an odd number of dots, we are left with a square of magnitude three on which the same procedure is repeated. This, when abstracted is the step case of the schematic proof, i.e., $\mathcal{A}(n)$. It is possible that the user when constructing the example-proof realizes that the pattern repeats itself after the first instance of an instantiated step case. This gives a potential to exploit the user's intuition about the repetition of a pattern in the proof – in DIAMOND this is captured by the "repeat..." feature.

The idea behind "repeat..." is that during the process of interactive construction of example-proofs the system records the operations carried out so far. If the user indicates that the sequence of operations applied on the diagram constitutes a pattern which needs to be repeated on the remaining diagram, this feature allows the user to instruct the system to automatically do so. For example, consider the example-proof trace for $n = 4$ given in Table 5. It is apparent that after carrying out the first section of the table: i.e., one lcut and three split_ends, the pattern repeats itself on the rest of the square. "repeat..." allows this repetition. However, the operations can only be applied as far as possible, depending on the magnitude of the diagram. For example, in the first application of "repeat..." it is not possible to apply three occurrences of split_ends, but rather only two. It is the role of the system to detect such constraints. If the pattern can be successfully repeated until the diagram is exhausted, then this pattern of operations indeed forms an instantiated step case of the example-proof, i.e., $\mathcal{A}(4)$. No further example-proof is needed for the abstraction mechanism. The first step of the abstraction algorithm given in §7.5 has been carried out by the user who provided the possible structure $\mathcal{A}$. The system can now make this step case general by first checking if the pattern can be repeated, and then by finding a dependency function for the general number of applications of operations in the step case of the schematic proof.

However, the problem remains when a pattern which constitutes an instance of a step case of a schematic proof is not apparent to the user. In a more complex example the recursiveness of the proof may not be obvious. For instance, it may not be clear to the user that two rectangles of magnitude 8×5 and 13×8 are instances of the same recursively defined general diagram, namely, the rectangles of magnitudes $Fib(6) \times Fib(5)$ and $Fib(7) \times Fib(6)$ respectively, where $Fib(x)$ is the x-th Fibonacci number. In general, these are two instances of a general rectangle of magnitude $Fib(n+1) \times Fib(n)$. We cannot expect that the user will always be able to detect a pattern in an example-proof. If this is so, DIAMOND needs two example-proofs to construct a general schematic proof.

7.9 Summary

Diagrammatic schematic proofs are constructed from examples of proofs for instances of a theorem. DIAMOND uses an inductive inference to abstract a general pattern from particular examples of proofs. The particular formalization of a schematic proof in DIAMOND is simple, consisting of a single recursive call. Some analysis of the theorems that DIAMOND is targeted for indicates that this is a heuristically adequate choice. The recursive program proof by instantiation uniformly constructs a proof of each instance of the premise.

DIAMOND's abstraction mechanism is capable of extracting a linear dependency between the number of applications of operations in the example-proof, and the particular instance under consideration. Although there may be proofs where this dependency is more complex, e.g., exponential, this too, turns out to be a heuristically adequate choice.

Some proofs are not defined for all consecutive numbers, but may have homogeneous proofs for only a subset of numbers, e.g., they may have one proof for even and a different proof for odd numbers. Such proofs contain case splits, and in order to have a complete proof, a schematic proof for each case has to be defined.

DIAMOND needs two example-proofs in order to be able to construct a schematic proof. However, there are abstraction mechanisms which can abstract from one example only. Such mechanisms typically require some form of background knowledge which is not available to DIAMOND.

All abstraction mechanisms, including the one in DIAMOND, make an educated guess about the abstraction of the pattern of reasoning. This guess can still be fallible. So, the schematic proof could still be incorrect. Therefore, DIAMOND needs to verify the correctness of the schematic proof, and how this is done is the topic of the next chapter.

8

The Verification of Schematic Proofs

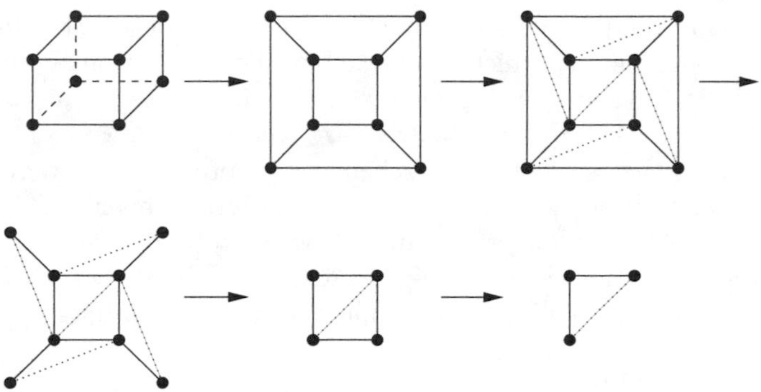

$$Vertices - Edges + Faces = 2$$

— AUGUSTIN LOUIS CAUCHY
in LAKATOS' *Proofs and Refutations*

The process of construction of diagrammatic proofs has been presented so far in two stages. The first stage is to prove diagrammatically ground instances of a conjecture at hand (Chapter 5). The second stage is to extract a proof structure common to these examples and capture this structure in a recursive program $\mathsf{proof}(n)$ called a schematic proof. The common structure is extracted using an abstraction algorithm (Chapter 7). The last stage is to prove the correctness of the induced schematic proof, i.e., we need to show that a schematic proof indeed proves the proposition for all n. This ensures that the transition from specific examples to a general proof is sound.

In this chapter we present a method which enables us to prove the correctness of schematic proofs for *particular* theorems. We present a theory of diagrams in which the verification is carried out. The idea is to show that when the geometric operations of a particular schematic proof arc applied to diagrams, they indeed result in the collection of correct diagrams which represent the theorem. We define in this chapter what is meant by a collection of correct diagrams. Furthermore, we define when a schematic proof is an algebraically correct proof of a theorem. A meta-level theorem which states when a particular object-level arithmetic conjecture *can* be diagrammatically proved using Diamond is needed in the end. It enables us to put all the pieces of information together, and show how theorems can be proved diagrammatically starting with a conjecture and finishing with a verified diagrammatic schematic proof of this conjecture.

In §8.1, we motivate the need for verification of schematic proofs and propose a theory of diagrams as a verification mechanism. Then, we present some of the primitives of the theory: the diagrams in §8.2, the operators in §8.3, the operations in §8.4, and finally in §8.5, the function definitions and lemmas which are needed for the verification of schematic proofs. In §8.6 we state the property of the correctness of a particular schematic proof. In §8.7 we define the size of a diagram, which is used to make explicit the link between a schematic proof and a theorem that it proves. In §8.8 we define and prove a desired general property of algebraic correctness of schematic proofs. In §8.9 we state and prove a theorem about the diagrammatic provability of an arithmetic conjecture. Finally, in §8.10 we discuss the implementation of our theory of diagrams.

8.1 Motivation

The motivation for defining a theory of diagrams is to verify the correctness of schematic proofs that Diamond constructs, because the example-proofs and their abstraction which forms a schematic proof are fallible.[50] The verification ensures that the transition from concreteness to generality of a diagrammatic proof is correct. In human reasoning this step is often omitted when humans are convinced that the examples used to induce a general schematic proof uniformly account for all cases of a theorem. This can sometimes result in erroneous proofs (see §4.5). In an automated reasoning system, however, we need to *formally* show the correctness of a schematic proof.

Diamond's abstraction mechanism is an inductive inference algorithm and thus an unproven, but informed *guess* of a general schematic proof. The requirement by the constructive ω-rule, given in Definition 1 (see §4.2), is that there is a uniform procedure which proves each premise. To ensure that the *guessed* schematic proof is a procedure which proves each premise, we need to show in some meta theory that $\mathsf{proof}(n)$ *uni-*

[50] The reader is referred to §7.3 for the formalization and representation of schematic proofs.

formly proves $P(n)$ for all n. A meta-level proof using diagrams of general magnitude would be an obvious method for verifying our schematic proofs. However, such meta-level proof reintroduces the need for manipulating abstraction devices (e.g., ellipsis) in diagrams which, as discussed in §3.4, we are trying to avoid.

One way of overcoming this problem is to define diagrams and operations in a theory of diagrams where we can express abstract diagrams symbolically rather than diagrammatically. In this theory we can verify schematic proofs by defining the notion of applicability of a posited proof. Given that a particular theorem is expressed as an equality, its schematic proof is correct if applying the operations specified in the schematic proof on the diagrammatic representation of the left hand side of the theorem results in the diagrammatic representation of the right hand side of the theorem. There are two conditions that need to be satisfied. The first condition is that there is an appropriate diagrammatic representation available for the mapping of the theorem into its diagrammatic representation. The second condition is that the operations of the schematic proof are defined on those diagrams. A verification proof is a meta-level proof, because it is a proof about a property of an object-level schematic proof.[51]

Before we can state the definition of the correctness property of schematic proofs, we need to formalize the machinery which will enable us to model the processes of a diagrammatic proof. Therefore, we need to formally define diagrams, operations on them, and the applicability of operations of a schematic proof.

8.2 Diagrams

Diagrams in the theory are defined to be of object type. Some examples of the different kinds of object names in the theory are: row, column, ell, frame, square, rectangle, and triangle.

Diagrams of the theory model natural numbers. DIAMOND's primitive notion of a concrete diagram, a dot, is represented in the theory as the natural number 1. Objects are introduced via a constructor function, diagram, which takes the name of the type of a diagram and the list of parameters for its magnitude. Thus, the type of constructor function diagram is name $\times$ nat list $\mapsto$ object.[52] So for instance, a square of magnitude 4 is expressed in the theory as diagram(square,[4]). All elementary and derived concrete diagrams are expressed using a primitive

[51]For the definitions of meta- and object-level proofs, see Glossary.

[52]Note that the data type nat stands for non-negative natural number of Peano arithmetic.

object dot, hence in the theory they can be expressed using a constructor function, the object name and some parameter representing a natural number for the magnitude of the diagram.

Constant $\emptyset$ denotes a null diagram, or in other words an empty diagram. We define that any diagram that is of 0 magnitude is an empty diagram (note that $a \in b$ denotes that a natural number a is an element of a list b; thus the type of an infix $\in$ is: $\mathsf{nat} \times \mathsf{nat\ list} \mapsto \mathsf{boolean}$):

$$0 \in s \rightarrow \mathsf{diagram}(x, s) \quad \equiv \quad \emptyset \tag{8.6}$$

Note also, that all triangles are equilateral (see §6.2). The reader is referred to §6.2 for a reminder of diagrams and their names. Here are some examples of diagrams:

$$\begin{gathered}
\mathsf{diagram}(\mathsf{row}, [n]) \\
\mathsf{diagram}(\mathsf{column}, [n]) \\
\mathsf{diagram}(\mathsf{ell}, [n]) \\
\mathsf{diagram}(\mathsf{square}, [n]) \\
\mathsf{diagram}(\mathsf{square}, [2n]) \\
\mathsf{diagram}(\mathsf{square}, [2n-1]) \\
\mathsf{diagram}(\mathsf{triangle}, [n]) \\
\mathsf{diagram}(\mathsf{rectangle}, [n, f(n)]) \\
\mathsf{diagram}(\mathsf{frame}, [n]) \\
\mathsf{diagram}(\mathsf{thick_frame}, [2n+1])
\end{gathered}$$

8.3 Operators

This section gives the operators available in the theory. First, we write diagrammatic equality using $\stackrel{d}{=}$ which denotes that two lists of diagrams are identical. Here is the definition of $\stackrel{d}{=}$:

$$\mathsf{X} \stackrel{d}{=} \mathsf{Y} \longleftrightarrow \forall \mathsf{d}.\ \mathsf{count}(\mathsf{d}, \mathsf{X}) \;=\; \mathsf{count}(\mathsf{d}, \mathsf{Y})$$

where the function count can be defined by:

$$\begin{aligned}
\mathsf{count}(\mathsf{d}, [\,]) &= 0 \\
\mathsf{count}(\mathsf{d}, \mathsf{d} :: \mathsf{D}) &= 1 + \mathsf{count}(\mathsf{d}, \mathsf{D}) \\
\mathsf{d} \neq \mathsf{e} \rightarrow \mathsf{count}(\mathsf{d}, \mathsf{e} :: \mathsf{D}) &= \mathsf{count}(\mathsf{d}, \mathsf{D})
\end{aligned}$$

Diagrammatic equality $\stackrel{d}{=}$ is a larger relation than an arithmetic equality $=$, because it has all the properties of $=$, i.e., reflexivity, symmetry, transitivity and substitution properties, plus an additional one: the order of elements in a list does not matter. Therefore, two lists of diagrams, X and Y, are diagrammatically equal, $\mathsf{X} \stackrel{d}{=} \mathsf{Y}$, even if the orders in which the diagrams are listed in both lists differ.[53]

We now define some operators that introduce the existence of several diagrams.

- @ – append on lists,
- :: and nil – list constructors (concatenation of elements onto a list, and an empty list),
- $\otimes$ – nat $\times$ object list $\mapsto$ object list (it is an infix operator which introduces a combination of a number of identical lists of diagrams),
- $\uplus$ – nat $\times$ nat $\times$ (nat $\mapsto$ object) $\mapsto$ object list (it denotes a collection of diagrams of increasing magnitudes which are all of the same kind; it is analogous to $\sum$ for summation of integers).

Here is the recursive definition of $\biguplus_{i=a}^{b}$ for all $a \leq b$:[54]

$$(8.7) \qquad \biguplus_{i=a}^{a} \mathsf{diagram}(\mathsf{name}, f(i)) \stackrel{d}{=} [\mathsf{diagram}(\mathsf{name}, f(a))]$$

$$(8.8) \quad a \leq b \rightarrow \biguplus_{i=a}^{b+1} \mathsf{diagram}(\mathsf{name}, f(i)) \stackrel{d}{=} \biguplus_{i=a}^{b} \mathsf{diagram}(\mathsf{name}, f(i))@ [\mathsf{diagram}(\mathsf{name}, f(b+1))]$$

Note that f is some function which generates a list of natural numbers for a given number i. This list denotes the parameters for the magnitude of a diagram. Note also that:

$$\biguplus_{i=a}^{b} \mathsf{diagram}(\mathsf{name}, f(i)) \stackrel{d}{=}$$
$$[\mathsf{diagram}(\mathsf{name}, f(a)), \mathsf{diagram}(\mathsf{name}, f(a+1)), \ldots, \mathsf{diagram}(\mathsf{name}, f(b))]$$

8.4 Operations

Diagrammatic operations are represented via a function op : opname $\times$ object list $\mapsto$ object list. Figure 36 and Figure 37 define some operations on diagrams. Note that it is also possible to define new operations in DIAMOND by adding them to the repertoire of operations available in the construction of example-proofs, and to the theory of diagrams, i.e., the verification mechanism. A diagrammatic operation is valid if it preserves the sum of natural numbers that the resulting diagrams represent.

[53]Note that our definition of diagrammatic equality of lists is equivalent to bag equality. The order of the elements in a bag (sometimes called multi-set) does not matter (Manna and Waldinger 1985).

[54]Note that to simplify the notation we write $\biguplus_{i=a}^{b} D(i)$ instead of $\uplus(a, b, \lambda i.D(i))$.

(8.9) $\mathsf{op}(\mathsf{lcut}, \mathsf{diagram}(\mathsf{square}, [n+1]) :: \mathsf{D}) \equiv$
$[\mathsf{diagram}(\mathsf{square}, [n]), \mathsf{diagram}(\mathsf{ell}, [n+1])]@\mathsf{D}$

(8.10) $\mathsf{op}(\mathsf{lcut}, \mathsf{diagram}(\mathsf{triangle}, [n+2]) :: \mathsf{D}) \equiv$
$[\mathsf{diagram}(\mathsf{triangle}, [n]), \mathsf{diagram}(\mathsf{ell}, [n+2])]@\mathsf{D}$

(8.11) $\mathsf{op}(\mathsf{split\_row}, \mathsf{diagram}(\mathsf{ell}, [n+1]) :: \mathsf{D}) \equiv$
$[\mathsf{diagram}(\mathsf{column}, [n]), \mathsf{diagram}(\mathsf{row}, [n+1])]@\mathsf{D}$

(8.12) $\mathsf{op}(\mathsf{split\_row}, \mathsf{diagram}(\mathsf{rectangle}, [n, n+1]) :: \mathsf{D}) \equiv$
$[\mathsf{diagram}(\mathsf{square}, [n]), \mathsf{diagram}(\mathsf{row}, [n])]@\mathsf{D}$

(8.13) $\mathsf{op}(\mathsf{split\_col}, \mathsf{diagram}(\mathsf{rectangle}, [n, f(n)+1]) :: \mathsf{D}) \equiv$
$[\mathsf{diagram}(\mathsf{rectangle}, [n, f(n)]), \mathsf{diagram}(\mathsf{row}, [n])]@\mathsf{D}$

(8.14) $\mathsf{op}(\mathsf{split\_col}, \mathsf{diagram}(\mathsf{square}, [n+1]) :: \mathsf{D}) \equiv$
$[\mathsf{diagram}(\mathsf{rectangle}, [n, n+1]), \mathsf{diagram}(\mathsf{column}, [n+1])]@\mathsf{D}$

(8.15) $\mathsf{op}(\mathsf{split\_col}, \mathsf{diagram}(\mathsf{rectangle}, [n+1, n]) :: \mathsf{D}) \equiv$
$[\mathsf{diagram}(\mathsf{square}, [n]), \mathsf{diagram}(\mathsf{column}, [n])]@\mathsf{D}$

(8.16) $\mathsf{op}(\mathsf{split\_col}, \mathsf{diagram}(\mathsf{rectangle}, [n+1, f(n+1)]) :: \mathsf{D}) \equiv$
$[\mathsf{diagram}(\mathsf{rectangle}, [n, f(n+1)]), \mathsf{diagram}(\mathsf{column}, [f(n+1)])]@\mathsf{D}$

(8.17) $\mathsf{op}(\mathsf{split\_diagonally}, \mathsf{diagram}(\mathsf{square}, [n+1]) :: \mathsf{D}) \equiv$
$[\mathsf{diagram}(\mathsf{triangle}, [n+1]), \mathsf{diagram}(\mathsf{triangle}, [n])]@\mathsf{D}$

(8.18) $\mathsf{op}(\mathsf{split\_diagonally}, \mathsf{diagram}(\mathsf{rectangle}, [n+1, n]) :: \mathsf{D}) \equiv$
$(2 \otimes [\mathsf{diagram}(\mathsf{triangle}, [n])])@\mathsf{D}$

(8.19) $\mathsf{op}(\mathsf{split\_diagonally}, \mathsf{diagram}(\mathsf{rectangle}, [n, n+1]) :: \mathsf{D}) \equiv$
$(2 \otimes [\mathsf{diagram}(\mathsf{triangle}, [n])])@\mathsf{D}$

(8.20) $\mathsf{op}(\mathsf{split\_outer\_frame}, \mathsf{diagram}(\mathsf{square}, [n+2]) :: \mathsf{D}) \equiv$
$[\mathsf{diagram}(\mathsf{square}, [n]), \mathsf{diagram}(\mathsf{frame}, [n+2])]@\mathsf{D}$

(8.21) $\mathsf{op}(\mathsf{split\_inner\_dot}, \mathsf{diagram}(\mathsf{square}, [2n+1]) :: \mathsf{D}) \equiv$
$[\mathsf{diagram}(\mathsf{thick\_frame}, [2n+1]), \mathsf{diagram}(\mathsf{square}, [1])]@\mathsf{D}$

(8.22) $\mathsf{op}(\mathsf{split2four}, \mathsf{diagram}(\mathsf{square}, [2n]) :: \mathsf{D}) \equiv$
$(4 \otimes [\mathsf{diagram}(\mathsf{square}, [n])])@\mathsf{D}$

(8.23) $\mathsf{op}(\mathsf{rotate90}, \mathsf{diagram}(\mathsf{rectangle}, [n, f(n)]) :: \mathsf{D}) \equiv$
$[\mathsf{diagram}(\mathsf{rectangle}, [f(n), n])]@\mathsf{D}$

(8.24) $\mathsf{op}(\mathsf{split\_sqr}, \mathsf{diagram}(\mathsf{rectangle}, [n+f(n), n]) :: \mathsf{D}) \equiv$
$[\mathsf{diagram}(\mathsf{rectangle}, [f(n), n]), \mathsf{diagram}(\mathsf{square}, [n])]@\mathsf{D}$

FIGURE 36 Diagrammatic operations in the theory of diagrams – part 1.

(8.25) $\mathsf{op}(\mathsf{split\_sqr}, \mathsf{diagram}(\mathsf{rectangle}, [n, n + f(n)]) :: \mathsf{D}) \equiv$
$[\mathsf{diagram}(\mathsf{rectangle}, [n, f(n)]), \mathsf{diagram}(\mathsf{square}, [n])]@\mathsf{D}$

(8.26) $\mathsf{op}(\mathsf{split\_side}, \mathsf{diagram}(\mathsf{triangle}, [n + 1]) :: \mathsf{D}) \equiv$
$[\mathsf{diagram}(\mathsf{triangle}, [n]), \mathsf{diagram}(\mathsf{row}, [n + 1])]@\mathsf{D}$

(8.27) $\mathsf{op}(\mathsf{split\_tst}, \mathsf{diagram}(\mathsf{triangle}, [2n]) :: \mathsf{D}) \equiv$
$((2 \otimes [\mathsf{diagram}(\mathsf{triangle}, [n])])@[\mathsf{diagram}(\mathsf{square}, [n])])@\mathsf{D}$

(8.28) $\mathsf{op}(\mathsf{split\_tst}, \mathsf{diagram}(\mathsf{triangle}, [2n + 1]) :: \mathsf{D}) \equiv$
$((2 \otimes \mathsf{diagram}(\mathsf{triangle}, [n])])@[\mathsf{diagram}(\mathsf{square}, [n + 1])])@\mathsf{D}$

(8.29) $\mathsf{op}(\mathsf{split\_dia\_ends}, \mathsf{diagram}(\mathsf{ell}, [n + 1]) :: \mathsf{D}) \equiv$
$[\mathsf{diagram}(\mathsf{ell}, [n]), \mathsf{diagram}(\mathsf{column}, [1]), \mathsf{diagram}(\mathsf{row}, [1])]@\mathsf{D}$

(8.30) $\mathsf{op}(\mathsf{split\_frame}, \mathsf{diagram}(\mathsf{frame}, [n + 1]) :: \mathsf{D}) \equiv$
$((2 \otimes [\mathsf{diagram}(\mathsf{row}, [n])])@(2 \otimes [\mathsf{diagram}(\mathsf{column}, [n])]))@\mathsf{D}$

(8.31) $\mathsf{op}(\mathsf{split\_tframe}, \mathsf{diagram}(\mathsf{thick\_frame}, [2n + 1]) :: \mathsf{D}) \equiv$
$((2 \otimes [\mathsf{diagram}(\mathsf{rectangle}, [n + 1, n])])@$
$(2 \otimes [\mathsf{diagram}(\mathsf{rectangle}, [n, n + 1])]))@\mathsf{D}$

FIGURE 37 Diagrammatic operations in the theory of diagrams – part 2.

8.5 Function definitions

8.5.1 One_Apply and Apply

Here we define what it means to apply an operation on a diagram several times. We use a function apply which is of the type apply: (opname × nat) list × object list ↦ object list, and a function one_apply which is of the type one_apply : nat × opname × object list ↦ object list. Recall that object list is the type of a list of diagrams. Let:

(8.32) $\mathsf{one\_apply}(0, \mathsf{opnm}, \mathsf{D}) \stackrel{d}{=} \mathsf{D}$

(8.33) $\mathsf{one\_apply}(n + 1, \mathsf{opnm}, \mathsf{D}) \stackrel{d}{=} \mathsf{op}(\mathsf{opnm}, \mathsf{one\_apply}(n, \mathsf{opnm}, \mathsf{D}))$

(8.34) $\mathsf{apply}([\,], \mathsf{D}) \stackrel{d}{=} \mathsf{D}$

(8.35) $\mathsf{apply}((\mathsf{opnm}, x) :: \mathsf{opss}, \mathsf{D}) \stackrel{d}{=} \mathsf{apply}(\mathsf{opss}, \mathsf{one\_apply}(x, \mathsf{opnm}, \mathsf{D})$

Note that opss is a list of pairs of an operation and the number of times that this operation is applied to a diagram.

8.5.2 Equations

Here we give an axiom about a null diagram defined in §8.2:

$$\emptyset :: \mathsf{D} \stackrel{d}{=} \mathsf{D} \tag{8.36}$$

Here are some theorems.

$$\mathsf{op}(\mathsf{opnm}, \mathsf{D} :: \mathsf{D}_s) \stackrel{d}{=} \mathsf{op}(\mathsf{opnm}, [\mathsf{D}])@\mathsf{D}_s \tag{8.37}$$

$$\mathsf{one\_apply}(n, \mathsf{opnm}, \mathsf{D} :: \mathsf{D}_s) \stackrel{d}{=} \mathsf{one\_apply}(n, [\mathsf{D}])@\mathsf{D}_s \tag{8.38}$$

$$\mathsf{apply}(\mathsf{ops}, \mathsf{D} :: \mathsf{D}_s) \stackrel{d}{=} \mathsf{apply}(\mathsf{ops}, [\mathsf{D}])@\mathsf{D}_s \tag{8.39}$$

Proof. The proof of equation (8.37) is carried out by a case analysis of the operations. We give here an example of one case. All other cases of defined operations on diagrams (i.e., (8.9) through to (8.31) in Figures 36 and 37) are similar. Let $\mathsf{opnm} = \mathsf{lcut}$ and $\mathsf{D} = \mathsf{diagram}(\mathsf{square}, [n+1])$ in $\mathsf{op}(\mathsf{opnm}, \mathsf{D} :: \mathsf{D}_s) \stackrel{d}{=} \mathsf{op}(\mathsf{opnm}, [\mathsf{D}])@\mathsf{D}_s$. Then we have:

$$\begin{array}{ccc} \mathsf{op}(\mathsf{lcut}, \mathsf{diagram}(\mathsf{square}, [n{+}1]) :: \mathsf{D}_s) & \stackrel{d}{=} & \mathsf{op}(\mathsf{lcut}, [\mathsf{diagram}(\mathsf{square}, [n{+}1])])@\mathsf{D}_s \\ (8.9) & \Downarrow & (8.9) \\ [\mathsf{diagram}(\mathsf{square}, [n]), & \stackrel{d}{=} & ([\mathsf{diagram}(\mathsf{square}, [n]), \\ \mathsf{diagram}(\mathsf{ell}, [n+1])]@\mathsf{D}_s & & \mathsf{diagram}(\mathsf{ell}, [n+1])]@[\,])@\mathsf{D}_s \end{array}$$

□

Proof. The proof of equation (8.38) is carried out by mathematical induction on n using the rules (8.32), (8.33) and (8.37).

Base case: $n = 0$

$$\begin{array}{ccc} \mathsf{one\_apply}(0, \mathsf{opnm}, \mathsf{D} :: \mathsf{D}_s) & \stackrel{d}{=} & \mathsf{one\_apply}(0, \mathsf{opnm}, [\mathsf{D}])@\mathsf{D}_s \\ (8.32) & \Downarrow & (8.32) \\ \mathsf{D} :: \mathsf{D}_s & \stackrel{d}{=} & [\mathsf{D}]@\mathsf{D}_s \end{array}$$

Step case:

Hypothesis: $\mathsf{one\_apply}(n, \mathsf{opnm}, \mathsf{D} :: \mathsf{D}_s) \stackrel{d}{=} \mathsf{one\_apply}(n, \mathsf{opnm}, [\mathsf{D}])@\mathsf{D}_s$
Conclusion:

$$\begin{array}{rcl}
\mathsf{one_apply}(n+1,\mathsf{opnm},\mathsf{D} :: \mathsf{D}_s) & \stackrel{d}{=} & \mathsf{one_apply}(n+1,\mathsf{opnm},[\mathsf{D}])@\mathsf{D}_s \\
(8.33) & \Downarrow & (8.33) \\
\mathsf{op}(\mathsf{opnm},\mathsf{one_apply}(n,\mathsf{opnm},\mathsf{D} :: \mathsf{D}_s)) & \stackrel{d}{=} & \mathsf{op}(\mathsf{opnm},\mathsf{one_apply}(n,\mathsf{opnm},\mathsf{D}]))@\mathsf{D}_s \\
\text{hypothesis} & \Downarrow & \\
\mathsf{op}(\mathsf{opnm},\mathsf{one_apply}(n,\mathsf{opnm},[\mathsf{D}])@\mathsf{D}_s) & \stackrel{d}{=} & \mathsf{op}(\mathsf{opnm},\mathsf{one_apply}(n,\mathsf{opnm},[\mathsf{D}]))@\mathsf{D}_s \\
\text{generalize:} & \Downarrow & \text{let } \mathsf{one_apply}(n,\mathsf{opnm},[\mathsf{D}]) = \mathsf{G} \\
\mathsf{op}(\mathsf{opnm},\mathsf{G}@\mathsf{D}_s) & \stackrel{d}{=} & \mathsf{op}(\mathsf{opnm},\mathsf{G})@\mathsf{D}_s
\end{array}$$

Note that the case where G=[ ], i.e., $\mathsf{one\_apply}(n,\mathsf{opnm},[\mathsf{D}]) = [\,]$ never arises, because an operation opnm is applied in $\mathsf{one\_apply}(n,\mathsf{opnm},[\mathsf{D}]) = \mathsf{G}$ to a non-empty diagram list [D], and all the operations preserve the natural number that a diagram represents, hence G cannot be empty either.

If $\mathsf{G} = \mathsf{G}_1 :: \mathsf{G}_s$ then $\mathsf{op}(\mathsf{opnm},(\mathsf{G}_1 :: \mathsf{G}_s)@\mathsf{D}_s) \stackrel{d}{=} \mathsf{op}(\mathsf{opnm},\mathsf{G}_1 :: \mathsf{G}_s)\ @\ \mathsf{D}_s$, which is true by applying (8.37) on both sides of the diagrammatic equality. □

Proof. The proof of equation (8.39) is carried out by mathematical induction on the list ops using the rules (8.34), (8.35) and (8.38).

Base case: $\mathsf{ops} = [\,]$

$$\begin{array}{rcl}
\mathsf{apply}([\,],\mathsf{D} :: \mathsf{D}_s) & \stackrel{d}{=} & \mathsf{apply}([\,],[\mathsf{D}])@\mathsf{D}_s \\
(8.34) & \Downarrow & (8.34) \\
\mathsf{D} :: \mathsf{D}_s & \stackrel{d}{=} & [\mathsf{D}]@\mathsf{D}_s
\end{array}$$

Step case:

Hypothesis: $\mathsf{apply}(\mathsf{ops},\mathsf{D} :: \mathsf{D}_s) \stackrel{d}{=} \mathsf{apply}(\mathsf{ops},[\mathsf{D}])@\mathsf{D}_s$
Conclusion:

$$\begin{array}{rcl}
\mathsf{apply}((\mathsf{opnm},n) :: \mathsf{ops},\mathsf{D} :: \mathsf{D}_s) & \stackrel{d}{=} & \mathsf{apply}((\mathsf{opnm},n) :: \mathsf{ops},[\mathsf{D}])@\mathsf{D}_s \\
(8.35) & \Downarrow & (8.35) \\
\mathsf{apply}(\mathsf{ops},\mathsf{one_apply}(n,\mathsf{opnm},\mathsf{D} :: \mathsf{D}_s)) & \stackrel{d}{=} & \mathsf{apply}(\mathsf{ops},\mathsf{one_apply}(n,\mathsf{opnm},[\mathsf{D}]))@\mathsf{D}_s \\
(8.38) & \Downarrow & \\
\mathsf{apply}(\mathsf{ops},\mathsf{one_apply}(n,\mathsf{opnm},[\mathsf{D}])@\mathsf{D}_s) & \stackrel{d}{=} & \mathsf{apply}(\mathsf{ops},\mathsf{one_apply}(n,\mathsf{opnm},[\mathsf{D}]))@\mathsf{D}_s \\
\text{generalize:} & \Downarrow & \text{let } \mathsf{one_apply}(n,\mathsf{opnm},[\mathsf{D}]) = \mathsf{G} \\
\mathsf{apply}(\mathsf{ops},\mathsf{G}@\mathsf{D}_s) & \stackrel{d}{=} & \mathsf{apply}(\mathsf{ops},\mathsf{G})@\mathsf{D}_s
\end{array}$$

Note that the case $\mathsf{G} = [\,]$ never arises for the same reasoning as in the proof of (8.38).

If $\mathsf{G}{=}\mathsf{G_1} :: \mathsf{G_s}$ then $\mathsf{apply(ops, (G_1 :: G_s)@D_{\it s})} \stackrel{d}{=} \mathsf{apply(ops, G_1 :: G_s)@D_{\it s}}$, which is true by appealing to the hypothesis on both sides of the diagrammatic equality. □

8.5.3 Mapping relation dmap

Let dmap denote a relation between a particular class of statements of arithmetic and their equivalent diagrammatic expressions in the theory of diagrams. The equivalence is defined to be over the *size* of the diagram. The size of a diagram is defined to be the number of counters (dots) in the diagram, i.e., the natural number that the diagram represents. dmap takes two arguments, an arithmetic expression and a list of diagrams which could collectively represent this expression. Hence, the type of the relation dmap is nat × object list. Here are some general mappings:

(8.40) $$\mathsf{dmap}(0, [\,])$$

(8.41) $$\mathsf{dmap}(n^2, [\mathsf{diagram}(\mathsf{square}, [n])])$$

(8.42) $$\mathsf{dmap}(2n-1, [\mathsf{diagram}(\mathsf{ell}, [n])])$$

(8.43) $$\mathsf{dmap}(n, [\mathsf{diagram}(\mathsf{row}, [n])])$$

(8.44) $$\mathsf{dmap}(n, [\mathsf{diagram}(\mathsf{column}, [n])])$$

(8.45) $$\mathsf{dmap}(n \times f(n), [\mathsf{diagram}(\mathsf{rectangle}, [n, f(n)])])$$

(8.46) $$\mathsf{dmap}(\tfrac{n(n+1)}{2}, [\mathsf{diagram}(\mathsf{triangle}, [n])])$$

(8.47) $$\mathsf{dmap}(4(n-1), [\mathsf{diagram}(\mathsf{frame}, [n])])$$

(8.48) $$\mathsf{dmap}((2n+1)^2-1, [\mathsf{diagram}(\mathsf{thick\_frame}, [2n+1])])$$

(8.49) $$m \neq 0 \rightarrow \mathsf{dmap}(n+m, \mathsf{D} :: \mathsf{E}) \text{ such that } \mathsf{dmap}(n, [\mathsf{D}]) \text{ and } \mathsf{dmap}(m, \mathsf{E})$$

(8.50) $$\mathsf{dmap}(\textstyle\sum_{j=a}^{b} f(j), \biguplus_{j=a}^{b} \mathsf{D}_j) \text{ such that } \forall j, a \leq j \leq b, \mathsf{dmap}(f(j), [\mathsf{D}_j])$$

8.6 Correctness of Schematic Proofs

We have now formalized enough machinery to be able to define the correctness property of a schematic proof.

Definition 3 (Correctness of Schematic Proofs)
proof is a correct schematic proof of a particular conjecture $\forall n \;\; L(n) = R(n)$ if for all n there exist two lists of diagrams D and E such that $\mathsf{dmap}(L(n), \mathsf{D})$ and $\mathsf{dmap}(R(n), \mathsf{E})$, and

$$\mathsf{apply}\,(\mathsf{proof}\,(n),\;\mathsf{D}) \;\stackrel{d}{=}\; \mathsf{E}$$

It is possible to prove the property in Definition 3 only if $L(n)$, $R(n)$ and proof are known, i.e., for a specific case of a conjecture and a schematic proof. Knowing $L(n)$ and $R(n)$ allows us to infer some mapping relations which specify two lists of diagrams D and E. This satisfies the first part of Definition 3. In the next section we prove the correctness of a schematic proof for a particular conjecture at hand.

8.6.1 Example of Correctness Proof for a Schematic Proof

Here we prove the property given in Definition 3 for an example of a schematic proof of a theorem about the *sum of odd naturals.* The theorem is stated as $n^2 = \sum_{i=0}^{n}(2i-1)$. The schematic proof of this theorem is given as:[55]

(8.51) $$\mathsf{proof}(0) = [\,]$$

(8.52) $$\mathsf{proof}(n+1) = [(\mathsf{lcut}, 1)],\ \mathsf{proof}(n)$$

Proof. The proof of correctness of this particular schematic proof requires mathematical induction on n. The base case for $n = 0$ is trivial, since by (8.40) no operations are applied to an empty diagram list which results in $[\,]$. We consider a step case of induction.

Step case:

Hypothesis: for n

Using (8.41) notice

$$\mathsf{dmap}(n^2, [\mathsf{diagram}(\mathsf{square}, [n])]).$$

Hence, let $\mathsf{D} = [\mathsf{diagram}(\mathsf{square}, [n])]$.

Using (8.50) and (8.42) notice

$$\mathsf{dmap}(\sum_{i=0}^{n}(2i-1), \biguplus_{i=0}^{n} \mathsf{diagram}(\mathsf{ell}, [i])).$$

Hence, let $\mathsf{E} = \biguplus_{i=0}^{n} \mathsf{diagram}(\mathsf{ell}, [i])$. Now, we can express the hypothesis as:

$$\mathsf{apply}(\mathsf{proof}(n), [\mathsf{diagram}(\mathsf{square}, [n])]) \stackrel{d}{=} \biguplus_{i=0}^{n} \mathsf{diagram}(\mathsf{ell}, [i]).$$

Conclusion: for $n+1$

Similarly to the hypothesis, D and E are mapped for $n+1$.

---

[55] For the brevity of presentation we take a simpler version of the schematic proof which does not include the operation **split_ends** – we can do so without loss, because we assume that as **ell** has previously been showed to represent odd numbers.

$$\mathsf{apply}(\mathsf{proof}(n+1), [\mathsf{diagram}(\mathsf{square}, [n+1])]) \stackrel{d}{=} \biguplus_{i=0}^{n+1} \mathsf{diagram}(\mathsf{ell}, [i])$$

$$\mathsf{proof}(n+1) = [(\mathsf{lcut}, 1)],\ \mathsf{proof}(n) \Downarrow$$

$$\mathsf{apply}(((\mathsf{lcut}, 1),\ \mathsf{proof}(n)), [\mathsf{diagram}(\mathsf{square}, [n+1])]) \stackrel{d}{=} \biguplus_{i=0}^{n+1} \mathsf{diagram}(\mathsf{ell}, [i])$$

$$(8.35) \Downarrow$$

$$\begin{array}{r}\mathsf{apply}(\mathsf{proof}(n), \mathsf{one\_apply}(1, \mathsf{lcut}, \\ [\mathsf{diagram}(\mathsf{square}, [n+1])]))\end{array} \stackrel{d}{=} \biguplus_{i=0}^{n+1} \mathsf{diagram}(\mathsf{ell}, [i])$$

$$(8.33) \Downarrow$$

$$\begin{array}{r}\mathsf{apply}(\mathsf{proof}(n), \mathsf{op}(\mathsf{lcut}, \mathsf{one\_apply}(0, \mathsf{lcut}, \\ [\mathsf{diagram}(\mathsf{square}, [n+1])])))\end{array} \stackrel{d}{=} \biguplus_{i=0}^{n+1} \mathsf{diagram}(\mathsf{ell}, [i])$$

$$(8.32) \Downarrow$$

$$\mathsf{apply}(\mathsf{proof}(n), \mathsf{op}(\mathsf{lcut}, [\mathsf{diagram}(\mathsf{square}, [n+1])])) \stackrel{d}{=} \biguplus_{i=0}^{n+1} \mathsf{diagram}(\mathsf{ell}, [i])$$

$$(8.9) \Downarrow$$

$$\begin{array}{r}\mathsf{apply}(\mathsf{proof}(n), [\mathsf{diagram}(\mathsf{square}, [n]), \\ \mathsf{diagram}(\mathsf{ell}, [n+1])])\end{array} \stackrel{d}{=} \biguplus_{i=0}^{n+1} \mathsf{diagram}(\mathsf{ell}, [i])$$

$$(8.39) \Downarrow$$

$$\begin{array}{r}\mathsf{apply}(\mathsf{proof}(n), [\mathsf{diagram}(\mathsf{square}, [n])])@ \\ [\mathsf{diagram}(\mathsf{ell}, [n+1])]\end{array} \stackrel{d}{=} \biguplus_{i=0}^{n+1} \mathsf{diagram}(\mathsf{ell}, [i])$$

$$(\text{RHS of hypothesis}) \Downarrow$$

$$\biguplus_{i=0}^{n} \mathsf{diagram}(\mathsf{ell}, [i])@[\mathsf{diagram}(\mathsf{ell}, [n+1])] \stackrel{d}{=} \biguplus_{i=0}^{n+1} \mathsf{diagram}(\mathsf{ell}, [i])$$

$$(8.8) \Downarrow$$

$$\biguplus_{i=0}^{n+1} \mathsf{diagram}(\mathsf{ell}, [i]) \stackrel{d}{=} \biguplus_{i=0}^{n+1} \mathsf{diagram}(\mathsf{ell}, [i])$$

$\square$

8.7 Size of Diagrams

Definition 3 makes no claims about the link between a schematic proof and the theoremhood of a conjecture $\forall n\ L(n) = R(n)$. We still need to disprove the possibility of a *correct* schematic proof of a *false* conjecture. To establish that the conjecture is true when proved by a schematic proof, an explicit algebraic link between them needs to be defined. We establish this link via the *size of diagrams*. We first define the size of a diagram, and later, in §8.8, we state the theorem about the algebraic correctness of a schematic proof for a given conjecture.

Let us denote the size of the diagram D by $|\,\mathsf{D}\,|$. Here is a definition for the size of a diagram:

Definition 4 (Size of Diagrams)
The size of a list of diagrams is equal to the value of the arithmetic expression that it represents: if $\mathsf{dmap}(e, \mathsf{D})$ then $|\,\mathsf{D}\,| = e$.

Note that the type of $|\ \ |$ is: $\mathsf{object\ list} \mapsto \mathsf{nat}$. Using the property of size defined in Definition 4 on formulae from (8.40) to (8.50), we have the following:

$$|\,[\,]\,| = 0 \tag{8.53}$$

$$|\,[\mathsf{diagram}(\mathsf{square}, [n])]\,| = n^2 \tag{8.54}$$

$$|\,[\mathsf{diagram}(\mathsf{ell}, [n])]\,| = 2n - 1 \tag{8.55}$$

$$|\,[\mathsf{diagram}(\mathsf{row}, [n])]\,| = n \tag{8.56}$$

$$|\,[\mathsf{diagram}(\mathsf{column}, [n])]\,| = n \tag{8.57}$$

$$|\,[\mathsf{diagram}(\mathsf{rectangle}, [n, f(n)])]\,| = n \times f(n) \tag{8.58}$$

$$|\,[\mathsf{diagram}(\mathsf{triangle}, [n])]\,| = \frac{n(n+1)}{2} \tag{8.59}$$

$$|\,[\mathsf{diagram}(\mathsf{frame}, [n])]\,| = 4(n-1) \tag{8.60}$$

$$|\,[\mathsf{diagram}(\mathsf{thick\_frame}, [2n+1])]\,| = (2n+1)^2 - 1 \tag{8.61}$$

$$|\,\mathsf{D} :: \mathsf{E}\,| = |\,[\mathsf{D}]\,| + |\,\mathsf{E}\,| \tag{8.62}$$

$$\left|\,\biguplus_{j=a}^{b} \mathsf{D}_j\,\right| = \sum_{j=a}^{b} |\,[\mathsf{D}_j]\,| \tag{8.63}$$

We state now a lemma about the equality of sizes of two diagrammatically equal object lists.

Lemma 3 (Equality of Size of Two Diagram Lists)
Two diagrammatically equal lists of diagrams have the same size.

$$\mathsf{D} \stackrel{d}{=} \mathsf{E} \rightarrow |\,\mathsf{D}\,| = |\,\mathsf{E}\,|$$

Proof. The proof of Lemma 3 is straightforward by mathematical induction on the structure of D:

Base case: $\mathsf{D} = [\,]$

$$[\,] \stackrel{d}{=} \mathsf{E} \longrightarrow |\,[\,]\,| = |\,\mathsf{E}\,|$$

$$\Downarrow \text{ by substitution property of } \stackrel{d}{=}$$

$$|\,[\,]\,| \;=\; |\,[\,]\,|$$

Step case:

Hypothesis: for $\mathsf{D} = \mathsf{B}$, so $\mathsf{B} \stackrel{d}{=} \mathsf{E} \rightarrow |\,\mathsf{B}\,| = |\,\mathsf{E}\,|$ where E is universally quantified.

Conclusion: for $\mathsf{D} = \mathsf{A} :: \mathsf{B}$

$$\mathsf{A} :: \mathsf{B} \stackrel{d}{=} \mathsf{E}' \longrightarrow |\,\mathsf{A} :: \mathsf{B}\,| = |\,\mathsf{E}'\,|$$

E' $\neq [\,]$ since it contains at least A, so suppose E" = A::F then E' $\stackrel{d}{=}$ E". $\Downarrow$

$$\mathsf{A} :: \mathsf{B} \stackrel{d}{=} \mathsf{E}'' \longrightarrow |\,\mathsf{A} :: \mathsf{B}\,| = |\,\mathsf{E}''\,|$$

by substitution property of $\stackrel{d}{=}$ $\Downarrow$

$$\mathsf{A} :: \mathsf{B} \stackrel{d}{=} \mathsf{A} :: \mathsf{F} \longrightarrow |\,\mathsf{A} :: \mathsf{B}\,| = |\,\mathsf{A} :: \mathsf{F}\,|$$

$$\Downarrow \; (8.62)$$

$$\mathsf{A} :: \mathsf{B} \stackrel{d}{=} \mathsf{A} :: \mathsf{F} \longrightarrow |\,[\mathsf{A}]\,| + |\,\mathsf{B}\,| = |\,[\mathsf{A}]\,| + |\,\mathsf{F}\,|$$

$$\mathsf{X} \stackrel{d}{=} \mathsf{Y} \rightarrow \mathsf{Z}::\mathsf{X} \stackrel{d}{=} \mathsf{Z}::\mathsf{Y} \;\Downarrow\; M = N \rightarrow K + M = K + N$$

$$\mathsf{B} \stackrel{d}{=} \mathsf{F} \longrightarrow |\,\mathsf{B}\,| = |\,\mathsf{F}\,|$$

unify in hypothesis E with F $\Downarrow$

true

□

Now, we state a lemma about the preservation of the size of the sum of all resulting diagrams when an operation is applied on a diagram. For all operations that were just introduced, the following holds:

Lemma 4 (Size Invariance Under One Operation) *The size of the result of applying an operation to some diagrams is the same as the size of the diagrams before the operation was applied. Let* D *be some diagrams such that* $\mathsf{dmap}(e, \mathsf{D})$ *then:*

$$|\,\mathsf{op}(\mathsf{opname}, \mathsf{D})\,| = |\,\mathsf{D}\,|.$$

Proof. **(Case analysis on operations)** The proof of Lemma 4 consists of a case analysis of operations and mappings of arithmetic expressions. The case analysis is given in Table 6.

TABLE 6: Case analysis of operations.

opname	R		
lcut	8.9	D	diagram(square,$[x+1]$)::D
		DS	[diagram(square,$[x]$), diagram(ell,$[x+1]$)]@D
		|D|	$(x+1)^2+e$
		|DS|	$x^2+(2(x+1)-1)+e$
lcut	8.10	D	diagram(triangle,$[x+2]$)::D
		DS	[diagram(triangle,$[x]$), diagram(ell,$[x+2]$)]@D
		|D|	$\frac{(x+2)(x+3)}{2}+e$
		|DS|	$\frac{x(x+1)}{2}+(2(x+2)-1)+e$
split_row	8.11	D	diagram(ell,$[x+1]$)::D
		DS	[diagram(column,$[x]$), diagram(row,$[x+1]$)]@D
		|D|	$2(x+1)-1+e$
		|DS|	$x+(x+1)+e$
split_row	8.12	D	diagram(rectangle,$[x,x+1]$)::D
		DS	[diagram(square,$[x]$), diagram(row,$[x]$)]@D
		|D|	$x(x+1)+e$
		|DS|	x^2+x+e
split_col	8.13	D	diagram(rectangle,$[x,f(x)+1]$)::D
		DS	[diagram(rectangle,$[x,f(x)]$), diagram(row,$[x]$)]@D
		|D|	$x(f(x)+1)+e$
		|DS|	$x(f(x))+x+e$
split_col	8.14	D	diagram(square,$[x+1]$)::D
		DS	[diagram(rectangle,$[x,x+1]$), diagram(column,$[x+1]$)]@D
		|D|	$(x+1)^2+e$
		|DS|	$x(x+1)+(x+1)+e$
split_col	8.15	D	diagram(rectangle,$[x+1,x]$)::D
		DS	[diagram(square,$[x]$), diagram(column,$[x]$)]@D
		|D|	$(x+1)x+e$
		|DS|	x^2+x+e
split_col	8.16	D	diagram(rectangle,$[x+1,f(x+1)]$)::D
		DS	[diagram(rectangle,$[x,f(x+1)]$), diagram(column,$[f(x+1)]$)]@D
		|D|	$(x+1)f(x+1)+e$
		|DS|	$xf(x+1)+f(x+1)+e$
split_diagonally	8.17	D	diagram(square,$[x+1]$)::D
		DS	[diagram(triangle,$[x+1]$), diagram(triangle,$[x]$)]@D
		|D|	$(x+1)^2+e$
		|DS|	$\frac{(x+1)(x+2)}{2}+\frac{x(x+1)}{2}+e$
split_diagonally	8.18	D	diagram(rectangle,$[x+1,x]$)::D
		DS	$2\otimes$ [diagram(triangle,$[x]$)]@D
		|D|	$(x+1)x+e$
		|DS|	$2\frac{x(x+1)}{2}+e$

split_diagonally	8.19	D	diagram(rectangle,[$x, x+1$])::D
		DS	2 ⊗ [diagram(triangle,[x])]@D
		|D|	$x(x+1)+e$
		|DS|	$2\frac{x(x+1)}{2}+e$
split_outer_frame	8.20	D	diagram(square,[$x+2$])::D
		DS	[diagram(square,[x]), diagram(frame,[$x+2$])]@D
		|D|	$(x+2)^2+e$
		|DS|	$x^2+(4(x+1))+e$
split_inner_dot	8.21	D	diagram(square,[$2x+1$])::D
		DS	[diagram(thick_frame,[$2x+1$]), diagram(square,[1])]@D
		|D|	$(2x+1)^2+e$
		|DS|	$(2x+1)^2-1+1^2+e$
split2four	8.22	D	diagram(square,[$2x$])::D
		DS	4 ⊗ [diagram(square,x)]@D
		|D|	$(2x)^2+e$
		|DS|	$4x^2+e$
rotate90	8.23	D	diagram(rectangle,[$x, f(x)$])::D
		DS	[diagram(rectangle,[$f(x), x$])]@D
		|D|	$xf(x)+e$
		|DS|	$f(x)x+e$
split_sqr	8.24	D	diagram(rectangle,[$x+f(x), x$])::D
		DS	[diagram(rectangle,[$f(x), x$]), diagram(square,[x])]@D
		|D|	$(x+f(x))x+e$
		|DS|	$f(x)x+x^2+e$
split_sqr	8.25	D	diagram(rectangle,[$x, x+f(x)$])::D
		DS	[diagram(rectangle,[$x, f(x)$]), diagram(square,[x])]D
		|D|	$x(x+y)+e$
		|DS|	$xy+x^2+e$
split_side	8.26	D	diagram(triangle,[$x+1$])::D
		DS	[diagram(triangle,[x]), diagram(row,[$x+1$])]@D
		|D|	$\frac{(x+1)(x+2)}{2}+e$
		|DS|	$\frac{x(x+1)}{2}+(x+1)+e$
split_tst	8.27	D	diagram(triangle,[$2x$])::D
		DS	(2 ⊗ [diagram(triangle,[x])]) @ [diagram(square,[x])]@D
		|D|	$\frac{2x(2x+1)}{2}+e$
		|DS|	$2\frac{x(x+1)}{2}+x^2+e$
split_tst	8.28	D	diagram(triangle,[$2x+1$])::D
		DS	(2⊗ [diagram(triangle,[x])]) @ [diagram(square,[$x+1$])]@D
		|D|	$\frac{(2x+1)(2x+2)}{2}+e$
		|DS|	$2\frac{x(x+1)}{2}+(x+1)^2+e$
split_dia_ends	8.29	D	diagram(ell,[$x+1$])::D
		DS	[diagram(ell,[x]), diagram(column,[1]), diagram(row,[1])]@D
		|D|	$2(x+1)-1+e$
		|DS|	$(2x-1)+1+1+e$

split_frame	8.30	D	diagram(frame,$[x+1]$)::D
		DS	(2 ⊗ [diagram(row,$[x]$)]) @ (2 ⊗ [diagram(column,$[x]$)])@D
		|D|	$4(x-1+1)+e$
		|DS|	$2x+2x+e$
split_tframe	8.31	D	diagram(thick_frame,$[2x+1]$)::D
		DS	(2⊗ [diagram(rectangle,$[x+1,x]$)]) @ (2 ⊗ [diagram(rectangle,$[x,x+1]$)])@D
		|D|	$(2x+1)^2-1^2+e$
		|DS|	$2((x+1)x)+2(x(x+1))+e$

The proof consists of the following steps:

$$|\,\mathsf{op}(\mathsf{opname},\mathsf{D})\,| = |\,\mathsf{D}\,|$$
$$\text{using rule } R \Downarrow$$
$$|\,\mathsf{DS}\,| = |\,\mathsf{D}\,|$$

where Table 6 provides all cases. In particular, column 1 gives all cases of opname. Column 2 gives corresponding rules R used in the rewrite of the proof. Column 3 consists of four entries: for each case the first entry in column 3 gives the corresponding cases of D, the second one gives the corresponding DS's, the third one gives $|\,\mathsf{D}\,|$, and the fourth one gives $|\,\mathsf{DS}\,|$. Note that the values in the last two entries of column 3 are calculated using the rules of size given in (8.53) through to (8.63). Also, let dmap(e,D), hence $|\,\mathsf{D}\,| = e$.

Finally, the rest of the proof calculates that the two values $|\,\mathsf{D}\,|$ and $|\,\mathsf{DS}\,|$ in column 3 of Table 6 are the same. Note that in the calculation we subtract e from both sides of the equality first. The reference to the right of the calculation corresponds to the second column R in Table 6.

$$\begin{aligned}
&(8.9) && x^2+(2(x+1)-1) = x^2+2x+2-1 = x^2+2x+1 \\
& && = (x+1)^2 \\
&(8.10) && \tfrac{x(x+1)}{2}+(2(x+2)-1) = \tfrac{x^2+x+2(2x+4-1)}{2} = \tfrac{x^2+5x+6}{2} \\
& && = \tfrac{(x+2)(x+3)}{2} \\
&(8.11) && x+(x+1) = 2x+1 = 2(x+1)-1 \\
&(8.12) && x^2+x = x(x+1) \\
&(8.13) && x(f(x))+x = x(f(x)+1) \\
&(8.14) && x(x+1)+(x+1) = x^2+x+x+1 = (x+1)^2 \\
&(8.15) && x^2+x = (x+1)x \\
&(8.16) && x(f(x+1))+f(x+1) = (x+1)f(x+1) \\
&(8.17) && \tfrac{(x+1)(x+2)}{2}+\tfrac{x(x+1)}{2} = \tfrac{x^2+3x+2+x^2+x}{2} = \tfrac{2x^2+4x+2}{2} \\
& && = (x+1)^2 \\
&(8.18) && 2\tfrac{x(x+1)}{2} = x(x+1) = (x+1)x
\end{aligned}$$

$$
\begin{array}{lrl}
(8.19) & 2\frac{x(x+1)}{2} & = x(x+1) \\
(8.20) & x^2 + (4(x+1)) & = x^2 + 4x + 4 = (x+2)^2 \\
(8.21) & (2x+1)^2 - 1 + 1^2 & = (2x+1)^2 \\
(8.22) & 4x^2 & = (2x)^2 \\
(8.23) & f(x)x & = xf(x) \\
(8.24) & f(x)x + x^2 & = x(f(x)+x) = (x+f(x))x \\
(8.25) & xf(x) + x^2 & = x(f(x)+x) = x(x+f(x)) \\
(8.26) & \frac{x(x+1)}{2} + (x+1) & = \frac{x^2+x+2x+2}{2} = \frac{(x+1)(x+2)}{2} \\
(8.27) & 2\frac{x(x+1)}{2} + x^2 & = \frac{2x^2+2x+2x^2}{2} = \frac{4x^2+2x}{2} = \frac{2x(2x+1)}{2} \\
(8.28) & 2\frac{x(x+1)}{2} + (x+1)^2 & = \frac{2x^2+2x+2x^2+4x+2}{2} = \frac{4x^2+6x+2}{2} \\
 & & = \frac{(2x+1)(2x+2)}{2} \\
(8.29) & (2x-1)+1+1 & = 2x+1 = 2(x+1)-1 \\
(8.30) & 2x+2x & = 4x \\
(8.31) & 2((x+1)x) + 2(x(x+1)) & = 2x^2 + 2x + 2x^2 + 2x = 4x^2 + 4x \\
 & & = 4x^2 + 4x + 1 - 1 = (2x+1)^2 - 1
\end{array}
$$

□

The immediate consequence of Lemma 4 is the preservation of size when an operation is applied multiple times to some diagram.

Lemma 5 (Size Invariance Under Multiple Applications of One Operation)
The size of the result of applying an operation to some diagrams multiple times is the same as the size of the diagrams before the operation was applied multiple times. Let D *be some diagrams such that* dmap(e, D) *then:*

$$|\,\mathsf{one\_apply}(\mathsf{n}, \mathsf{opname}, \mathsf{D})\,| = |\,\mathsf{D}\,|$$

Proof. The proof of Lemma 5 is trivial by mathematical induction on n, using the rules (8.32) and (8.33) for the recursive definition of one_apply, and Lemma 4.

□

The immediate consequence of Lemma 4 and Lemma 5 is the preservation of size when several operations are applied multiple times to some diagram.

Lemma 6 (Size Invariance Under Multiple Operations)
The size of the result of applying a list of operations to some diagrams is the same as the size of the diagrams before the list of operations was applied. Let D *be some diagrams such that* dmap(e, D) *then:*

$$|\,\mathsf{apply}(\mathsf{ops}, \mathsf{D})\,| = |\,\mathsf{D}\,|$$

Proof. The proof of Lemma 6 is by straightforward mathematical induction on the structure of list ops, using the rules (8.34) and (8.35) for the recursive definition of apply, and Lemma 5.

□

8.8 Algebraic Correctness of Schematic Proofs

Apart from being diagrammatically correct, we want every schematic proof to be *algebraically correct* as well. A schematic proof is algebraically correct if the sizes of the diagrams representing both sides of the proposition after the operations of the schematic proof have been applied are the same. Theorem 7 states the property of algebraic correctness for any schematic proof.

Theorem 7 (Algebraic Correctness of Schematic Proofs)
For all instances of a schematic proof P *and for all pairs of lists of diagrams* D *and* E*, a schematic proof* P *is algebraically correct, that is,*

$$\mathsf{apply}\,(\mathsf{P},\mathsf{D}) \stackrel{d}{=} \mathsf{E} \longrightarrow |\,\mathsf{D}\,| \;=\; |\,\mathsf{E}\,|$$

Proof. The proof of Theorem 7 is straightforward by appealing to Lemma 3 and Lemma 6.

$$\begin{array}{c}
\mathsf{apply}(\mathsf{P},\mathsf{D}) \stackrel{d}{=} \mathsf{E} \longrightarrow |\,\mathsf{D}\,| = |\,\mathsf{E}\,| \\
\text{by Lemma 3} \Downarrow \\
|\,\mathsf{apply}(\mathsf{P},\mathsf{D})\,| = |\,\mathsf{E}\,| \longrightarrow |\,\mathsf{D}\,| = |\,\mathsf{E}\,| \\
\text{by Lemma 6} \Downarrow \\
|\,\mathsf{D}\,| = |\,\mathsf{E}\,| \longrightarrow |\,\mathsf{D}\,| = |\,\mathsf{E}\,|
\end{array}$$

□

8.9 Arithmetic Conjecture and Diagrammatic Proof

There is one last theorem needed in the formalization of diagrammatic theory which will allow us *to prove* theorems of arithmetic using *diagrammatic* proofs. We state in Theorem 8 the property about the diagrammatic provability of arithmetic arguments.

Theorem 8 (Diagrammatic Provability of an Arithmetic Conjecture)
A conjecture $\forall n\ L(n) = R(n)$ *is diagrammatically provable if and only if for all* n *there exist two lists of diagrams* D *and* E *such that* $\mathsf{dmap}(L(n),\mathsf{D})$ *and* $\mathsf{dmap}(R(n),\mathsf{E})$*, and*

$$|\,\mathsf{D}\,| = |\,\mathsf{E}\,| \longrightarrow L(n) = R(n)$$

Proof. The proof of Theorem 8 is trivial by the definition of size of a list of diagrams given in Definition 4.

□

8.9.1 Diagrammatic Provability for an Example

We consider now an example of an arithmetic conjecture and prove it diagrammatically using a schematic proof that DIAMOND constructs. Let the arithmetic conjecture be

$$\forall n.n^2 = \sum_{i=0}^{n}(2i-1)$$

and the schematic proof proof that DIAMOND constructed be as defined in (8.51) and (8.52).

Proof. Here are the reasoning steps of the proof:

1. Appealing to Theorem 8 we can discharge the conjecture by:
   - using (8.41) notice
     $$\mathsf{dmap}(n^2, [\mathsf{diagram}(\mathsf{square}, [n])]),$$
     hence let
     $$\mathsf{D} = [\mathsf{diagram}(\mathsf{square}, [n])],$$
   - using (8.50) and (8.42) notice
     $$\mathsf{dmap}(\sum_{i=0}^{n}(2i-1), \biguplus_{i=0}^{n} \mathsf{diagram}(\mathsf{ell}, [i])),$$
     hence let
     $$\mathsf{E} = \biguplus_{i=0}^{n} \mathsf{diagram}(\mathsf{ell}, [i]),$$

   and proving for all n

   $$|\,[\mathsf{diagram}(\mathsf{square}, [n])]\,| = \left|\biguplus_{i=0}^{n} \mathsf{diagram}(\mathsf{ell}, [i])\right| \tag{8.64}$$

2. Appealing to Theorem 7 and $\mathsf{proof}(n)$ that DIAMOND constructed, we can discharge the expression in (8.64) by proving for all n

   $$\mathsf{apply}\,(\mathsf{proof}(n), [\mathsf{diagram}(\mathsf{square}, [n])]) \stackrel{d}{=} \biguplus_{i=0}^{n} \mathsf{diagram}(\mathsf{ell}, [i]) \tag{8.65}$$

3. Finally, notice that we already proved (8.65) in §8.6.1.

□

8.10 Implementation of Theory of Diagrams

The verification mechanism that we formalized in this chapter is implemented in DIAMOND using *Clam*. *Clam* is a proof planner developed in Edinburgh (Bundy et al. 1991). It searches for a proof plan of a theorem. A proof plan is a high-level proof specification consisting of methods and strategies which specify clusters of inference rules that need to be applied in the object-level proof. An object-level verification proof can be obtained by executing the *Clam* proof plan in *Oyster* proof development system (Bundy et al. 1990). We are not interested in the intricacies of the object-level verification proof. Rather, we check if the verification theorem for our schematic proof is true by finding a proof plan. Hence, for the purposes of DIAMOND we do not execute a proof plan to obtain the object-level verification proof.

DIAMOND and *Clam* are linked together: a *Clam* server sits on top of DIAMOND and waits for *Clam* commands which are passed to it from DIAMOND.[56]

The implemented verification mechanism checks for the correctness of a constructed schematic proof, i.e., DIAMOND automatically checks whether the property given in Definition 3 is satisfied for a particular schematic proof. We follow the reasoning described in the previous section whereby the diagrammatic provability (Theorem 8) and algebraic correctness (Theorem 7) are used to discharge the original conjecture, which leaves us with the need to prove the correctness property of a schematic proof. The user can access the verification mechanism to check the correctness of a particular schematic proof via a command available on one of the menus in the main window of DIAMOND's graphical interface (see §5.6).

To automate the verification mechanism, all the primitives of the theory need to be loaded into *Clam* at the start of a DIAMOND session, i.e., the definitions of diagrams given in §8.2, the operators defined in §8.3, the operations defined in §8.4, and axioms, theorems and function definitions given in §8.5. Considering Definition 3, the following pieces of information need to be provided when a schematic proof is to be verified:

- the conjecture $\forall n.L(n) = R(n)$,
- the cases of the dmap relation which specify the two lists of diagrams D and E for $L(n)$ and $R(n)$ respectively,
- the schematic proof proof,
- the verification theorem $\forall n.\mathsf{apply}(\mathsf{proof}(n), \mathsf{D}) \stackrel{d}{=} \mathsf{E}$ with instantiated D and E.

[56] A separate window displays all the information that is passed to and is produced by *Clam*.

Therefore, the user is required to input to DIAMOND the theorem of natural number arithmetic expressed in a symbolic representation (see §8.10.2). DIAMOND tries to map the theorem using the relation dmap as defined in §8.5.3 into its diagrammatic representation to find D and E. The schematic proof is translated into the syntax of *Clam*, and loaded from DIAMOND into *Clam* as the definition of proof. DIAMOND formalizes the verification theorem using the provided information, and passes it to *Clam*. This completes the loading of all the definitions necessary for the correctness proof. Finally, DIAMOND initializes *Clam* which in turn starts searching for a proof plan of the verification theorem. If a proof plan can be found, then *Clam* passes it to DIAMOND to display it and to inform the user that the schematic proof is correct. If such a proof plan cannot be found, then the schematic proof may or may not be correct.

An interesting case to investigate would be a successful construction of a schematic proof of a theorem, but verifying the schematic proof in DIAMOND's theory of diagrams shows it is incorrect.[57] We are not interested in a trivial case of a theorem for which there is no mapping to a diagrammatic representation, or where the operations are not defined, so they cannot be used. We are interested in a theorem for which DIAMOND finds a schematic proof, but the verification shows that the schematic proof is incorrect. An example of a false schematic proof is the diagrammatic proof of *Euler's theorem* about polyhedra given at the beginning of this chapter. Check §A.5 for an explanation of a diagrammatic proof, as given by Cauchy (see Lakatos 1976). Although we cannot prove this theorem using DIAMOND, we can construct, as discussed in §4.5, a schematic proof, i.e., a uniform procedure that proves instances of this theorem. This proof satisfied human mathematician for a while, but it turns out that it is false, because not all of the examples of polyhedra were considered. Lakatos (1976, page 118) gives in the end a correct logical proof of this theorem which he attributes to Poincaré (1899). It would be interesting to identify other schematic proofs that human mathematicians found, but did not verify. Attempting to verifying such schematic proofs could potentially reveal that they are false.

8.10.1 Loaded Definitions and Lemmas

When a DIAMOND session is compiled, a *Clam* session is started as well, whereby all the definitions for diagrams, operators, operations, functions and axioms are loaded — equations from (8.6) to (8.36). The lemmas that we load are the theorems (8.37), (8.38) and (8.39) that were proved in §8.5.2.

[57]Note that our implementation of verification mechanism in *Clam* does *not* show that a verification theorem is false, it can only fail to find a proof plan.

The *Clam* proof methods and strategies which are available in the search of a proof plan include apply lemma, base case, induction strategy and normalization, plus all the methods loaded to *Clam* by default. We use depth-first proof planning search.

8.10.2 Theorem Mapping

We require that a theorem of natural number arithmetic, which is proved diagrammatically, is expressed as an equality with one universally quantified variable n, i.e., of the form $\forall n.L(n) = R(n)$. The user is required to enter this theorem using the appropriate syntax. Here is the grammar for this syntax:

$$
\begin{aligned}
term \equiv\ & term = term \\
\mid\ & term + term \\
\mid\ & term - term \\
\mid\ & term/term \\
\mid\ & term * term \\
\mid\ & sqr(term) \\
\mid\ & sum(term, term, \lambda(term, term)) \\
\mid\ & string \\
\mid\ & nat
\end{aligned}
$$

Note that in $sum(term, term, \lambda(term, term))$ the first argument is normally a natural number, the second argument is a variable, and the third argument is a lambda expression. The theory of diagrams is implemented over Peano natural numbers.

By Definition 3 it is required that there is a mapping of $L(n)$ to a diagrammatic representation D, and $R(n)$ to E. DIAMOND implements dmap as it is expressed in §8.5.3 in relations from (8.40) to (8.50), and searches for a mapping of $L(n)$ and $R(n)$. If no such mapping exists then the schematic proof cannot be verified.

8.10.3 Schematic Proof Encoding

Every time the user wants to verify a new schematic proof, a new recursive definition for this particular schematic proof has to be added to the verification mechanism. DIAMOND has built in functions which add new definitions to the implementation of theory of diagrams in *Clam*. A schematic proof can either be defined recursively or non-recursively. If the step case of the proof is empty,[58] and the base case consists of

[58]The reader is referred to §7.3 for a reminder of a formalization of a schematic proof.

operations which are parametrized over n, then the schematic proof is defined non-recursively as $\forall n.\mathsf{proof}(n) = \mathsf{ops}$.

8.10.4 Proof Plan

The proof plan for the verification of a schematic proof for the *sum of odd naturals* consists of the following methods: induction, step case and base case. The proof plan that *Clam* finds and passes back to Diamond looks as follows:

```
/* This is the pretty-printed form
   induction([(n:nat)-s(v0)])
        [base_case,
         step_case] then
   base_case(...)
*/
```

Note that the base case method in the proof plan consists of simple symbolic evaluation and rewriting. Besides base cases of inductive proofs it is also often used in non-inductive proofs. The object-level verification proof is given in §8.6.1. Its construction using the proof plan has not been automated, because it is not central to the ideas presented here.

8.11 Summary

The last stage of the procedure for construction of diagrammatic proofs as presented in §4.7 is to check the correctness of the abstracted schematic proof. A schematic proof is correct if it proves all cases (i.e., for all n) of the proposition. We formalized a theory of diagrams which allows us to check this correctness for a particular schematic proof. The language and the rules of the theory enable us to define the notion of applicability, and the correctness property of schematic proofs.

The algebraic correctness of a schematic proof ties the original theorem of natural number arithmetic to the diagrammatic schematic proof, and ensures that if the schematic proof of a diagrammatically expressed theorem is correct, then the corresponding statement of arithmetic is a theorem. The link between a diagrammatic theory and the theory of natural number arithmetic is the size of a diagram, which selects the natural number that the diagram represents. This number is the number of dots in the diagram.

The theorem regarding the diagrammatic provability of an arithmetic conjecture is used to show that a particular theorem of natural number arithmetic is diagrammatically provable using the diagrams available in

Diamond. If a constructed schematic proof is found to be correct, then the theorem of algebraic correctness and the theorem of diagrammatic provability can be used to formally justify why a schematic proof is a correct diagrammatic proof of an arithmetic theorem.

Diamond uses a proof planner *Clam* to implement the language and the rules of our diagrammatic theory, and to search for the proof of correctness of a schematic proof. Diamond finds a mapping of a theorem of arithmetic to its diagrammatic representation, and passes it along with the corresponding schematic proof to *Clam* to find a proof plan. If such a proof plan exists, then the schematic proof is correct. If it does not, then we cannot draw any conclusions about the correctness of the schematic proof.

This completes the three-stage process of constructing a formal diagrammatic proof of a theorem in Diamond. The fact that Diamond guarantees that the diagrammatic proof is correct and hence, that it is a formal proof of a theorem, is a powerful and novel component of a diagrammatic reasoning system. The existing reasoning systems either, unlike Diamond, construct *symbolic proofs* of diagrammatic situations which are guaranteed to be correct (e.g., see Shin's (1995) and Hammer's (1995) work), or they construct diagrammatic proofs which are, unlike in Diamond, *not guaranteed* to be correct (e.g., see Archimedes (Lindsay 1998) and IDR (Anderson and McCartney 1995)).

9

DIAMOND in Action

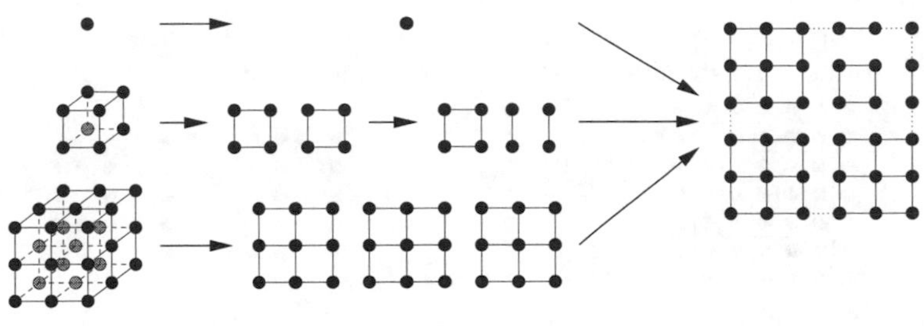

$$1^3 + 2^3 + 3^3 + \cdots + n^3 = (1 + 2 + 3 + \cdots + n)^2$$

— ALAN L. FRY
in NELSEN'S *Proofs Without Words*

In this book we investigated the use of diagrams in proofs of mathematical theorems. The theoretical ideas presented so far have been realized in the implementation of a diagrammatic reasoning system DIAMOND. In this chapter we present DIAMOND in action: in the process of proving some theorems we evaluate the ideas discussed in this book.

We begin in §9.1 with a detailed account of DIAMOND's construction of a diagrammatic proof, from example-proofs to a verified schematic proof. We continue in §9.2 by identifying the issues relevant for evaluating DIAMOND. This includes discussing DIAMOND's limitations in §9.2.4.

9.1 Example of DIAMOND's Diagrammatic Proof

We now present a DIAMOND session whereby a diagrammatic proof is constructed interactively with the user. The theorem under consideration is $(2n+1)^2 = 1 + 4(\sum_{i=0}^{n} 2i)$. The user, i.e., we are expected to provide examples of a proof, hence we give a possible diagrammatic representation of the theorem: $(2n+1)^2$ can be represented as a square of magnitude $2n+1$ for some particular n. Considering the right hand side of the theorem, $2i$ can be represented as a row of magnitude $2i$.

Multiplying this by 4 means that we have four rows. Here is the main idea for the schematic proof: it consists of splitting a square into frames, and then for each frame we split it into rows and columns (note that rows are the same as columns in terms of which natural number they represent). If the magnitude of a square is $2n + 1$ then one row will be of magnitude $2n$.

9.1.1 Constructing Example-Proof

Figure 38 shows an example-proof for the theorem under consideration where the parameter n is instantiated to 3. The user provides DIAMOND

1. Split a square of magnitude 7 (i.e., $2 \times 3 + 1$) three times into frames. This results in three frames and a dot.

2. For each frame, split it into rows and columns.

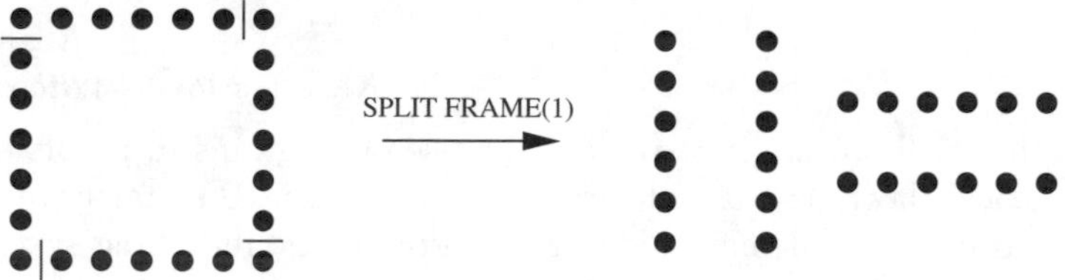

FIGURE 38 $(2n+1)^2 = 1+(4(2\times1)+4(2\times2)+\cdots+4(2n)) = 1 + 4(\sum_{i=0}^{n} 2i)$

with the values of the parameter n. For each ground instance of n, a diagrammatic proof is constructed. Figure 39 shows a screen shot of DIAMOND after the example-proof has been constructed. The proof trace of an example-proof for $n = 3$ consists of the following operations:

$$\begin{aligned}\mathsf{proof}(3) = {} & [(\mathsf{split\_outer\_frame}, 1), (\mathsf{split\_frame}, 1),\\ & (\mathsf{split\_outer\_frame}, 1), (\mathsf{split\_frame}, 1),\\ & (\mathsf{split\_outer\_frame}, 1), (\mathsf{split\_frame}, 1)]\end{aligned}$$

Figure 40 shows a screen shot of the trace window for this example. Another example-proof is constructed by the user for $n = 4$ and its

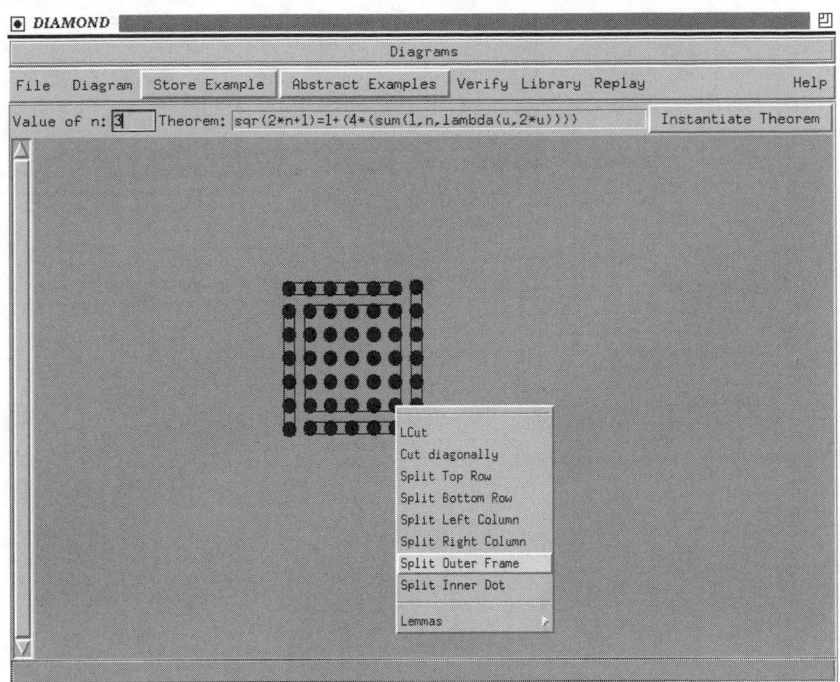

FIGURE 39 Screen shot of DIAMOND demonstrating an example-proof.

proof trace consists of the following operations:

$$\begin{aligned}\mathsf{proof}(4) = {} & [(\mathsf{split\_outer\_frame}, 1), (\mathsf{split\_frame}, 1),\\ & (\mathsf{split\_outer\_frame}, 1), (\mathsf{split\_frame}, 1),\\ & (\mathsf{split\_outer\_frame}, 1), (\mathsf{split\_frame}, 1),\\ & (\mathsf{split\_outer\_frame}, 1), (\mathsf{split\_frame}, 1)]\end{aligned}$$

9.1.2 DIAMOND's Schematic Proof

The number of inference steps in the proof of the theorem $(2n + 1)^2 = 1 + 4(\sum_{i=0}^{n} 2i)$, for which we showed an example-proof in Figure 38, depends on the parameter n. This means that the schematic proof of this theorem is defined recursively. The step case of the proof consists of two operations — split_outer_frame and split_frame:

(9.66) $$\begin{aligned}\mathsf{proof}(n+1) = {} & [(\mathsf{split\_outer\_frame}, 1), (\mathsf{split\_frame}, 1)],\\ & \mathsf{proof}(n)\end{aligned}$$

(9.67) $$\mathsf{proof}(1) = [\,]$$

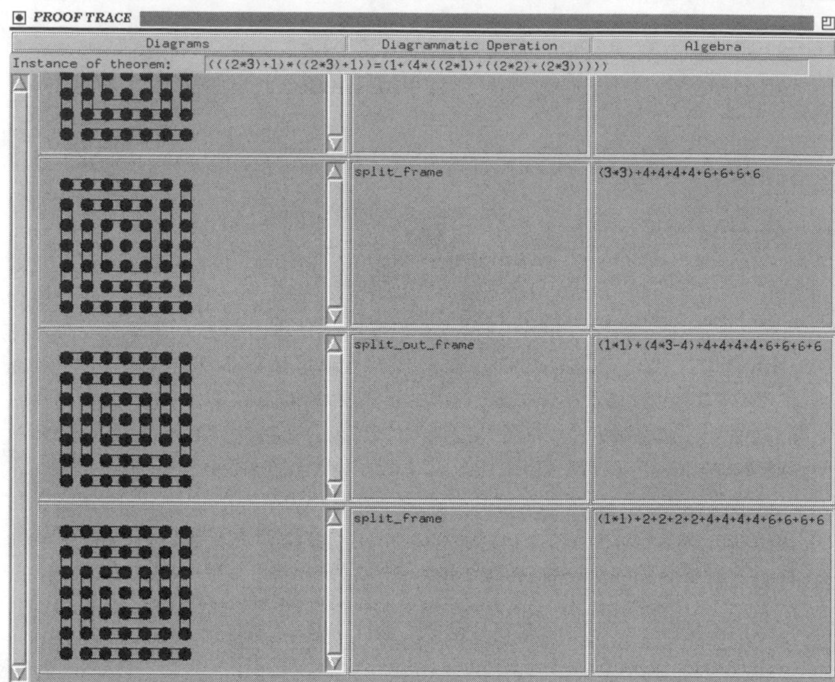

FIGURE 40 Screen shot of Diamond's trace for example-proof.

Diamond automatically abstracts the schematic proof formalized in (9.66) and (9.67) from two example-proofs given for $n = 3$ and $n = 4$.

We tested Diamond's abstraction mechanism for its robustness on two other sets of example-proofs, i.e., for $n = 3$ and $n = 5$, and for $n = 3$ and $n = 9$. Diamond abstracted the same schematic proof for both sets of example-proofs as shown above. This indicates that the abstraction mechanism is robust, i.e., it behaves as expected, irrespective of the different cases for which examples are given. The same test has been successfully carried out for all other theorems that were proved diagrammatically with Diamond.

9.1.3 Diamond's Verification Proof

The schematic proof that Diamond found and which was presented in §9.1.2 is automatically verified in *Clam* and found to be correct. Using (8.41) Diamond maps the left hand side of the equation expressing the theorem to $\mathsf{diagram}(\mathsf{square}, [2n+1])$, and using (8.41), (8.43), (8.44) and (8.50) Diamond can map the right hand side of the equation of the theorem to $\mathsf{diagram}(\mathsf{square}, [1]) :: (4 \otimes \biguplus_{j=0}^{n} \mathsf{diagram}(\mathsf{row}, [2j]))$. Notice that there are other possibilities for the mapping of the theorem. The theorem which *Clam* verifies is therefore stated as:

$$\forall n.\ \mathsf{apply}(\mathsf{proof}(n), [\mathsf{diagram}(\mathsf{square}, [2n+1])]) \stackrel{d}{=} \mathsf{diagram}(\mathsf{square}, [1]) :: (4 \otimes \biguplus_{j=0}^{n} \mathsf{diagram}(\mathsf{row}, [2j]))$$

Note that $\mathsf{proof}(n)$ is defined by (9.66) and (9.67). DIAMOND passes the recursive definition of $\mathsf{proof}(n)$ to *Clam*. *Clam* finds a proof plan for this theorem which consists of using an *induction strategy* on the universally quantified variable n, *step case* method, followed by a *base case* method that consists of symbolic evaluation which rewrites both sides of the equation to reach equality using various rewrite rules of the theory (see Chapter 8).

```
/* This is the pretty-printed form
   induction([(n:pnat)-s(v0)])
        [base_case,
         step_case] then
     base_case(...)
*/
```

The step case of the inductive proof is carried out by rippling, which uses annotations to guide rewriting.[59]

Proof. We give here just an outline of the object-level verification proof. The base case for $n = 0$ of the induction strategy is trivial. No operations are applied. After some symbolic evaluation both sides of the equation are equal to a list of one diagram, namely a square of magnitude 1. The hypothesis for n of the step case in the induction strategy is given above as the verification theorem. The outline of the proof looks as follows:

- Conclusion:

$$\mathsf{apply}(\mathsf{proof}(n+1), [\mathsf{diagram}(\mathsf{square}, [2(n+1)+1])]) \stackrel{d}{=} \mathsf{diagram}(\mathsf{square}, [1]) :: (4 \otimes \biguplus_{j=0}^{n+1} \mathsf{diagram}(\mathsf{row}, [2j]))$$

- Using (9.66), (8.20), (8.30), the definitions of apply (8.34), (8.35), and one_apply (8.32), (8.33), and the function which picks the right

[59] For more information on rippling, the reader is referred to Bundy et al. 1993.

diagram from the list of diagram, we have:

$$\begin{gathered}\mathsf{apply}(\mathsf{proof}(n), \mathsf{diagram}(\mathsf{square}, [2n+1]) :: \\ ((2 \otimes [\mathsf{diagram}(\mathsf{row}, [2(n+1)])])@ \\ (2 \otimes [\mathsf{diagram}(\mathsf{column}, [2(n+1)])]))) \\ \stackrel{d}{=} \\ \mathsf{diagram}(\mathsf{square}, [1]) :: (4 \otimes \biguplus_{j=0}^{n+1} \mathsf{diagram}(\mathsf{row}, [2j]))\end{gathered}$$

- Using a theorem that a column equals to a row, in addition to the definition of $\otimes$ we have:

$$\begin{gathered}\mathsf{apply}(\mathsf{proof}(n), \mathsf{diagram}(\mathsf{square}, [2n+1]) :: \\ (4 \otimes [\mathsf{diagram}(\mathsf{row}, [2(n+1)])])) \\ \stackrel{d}{=} \\ \mathsf{diagram}(\mathsf{square}, [1]) :: (4 \otimes \biguplus_{j=0}^{n+1} \mathsf{diagram}(\mathsf{row}, [2j]))\end{gathered}$$

- Using (8.39) we have:

$$\begin{gathered}\mathsf{apply}(\mathsf{proof}(n), [\mathsf{diagram}(\mathsf{square}, [2n+1])])@ \\ (4 \otimes [\mathsf{diagram}(\mathsf{row}, [2(n+1)])])) \\ \stackrel{d}{=} \\ \mathsf{diagram}(\mathsf{square}, [1]) :: (4 \otimes \biguplus_{j=0}^{n+1} \mathsf{diagram}(\mathsf{row}, [2j]))\end{gathered}$$

- Using the RHS of the hypothesis, (8.8) and definition of $\otimes$ we have:

$$\begin{gathered}\mathsf{diagram}(\mathsf{square}, [1]) :: (4 \otimes \biguplus_{j=0}^{n+1} \mathsf{diagram}(\mathsf{row}, [2j])) \\ \stackrel{d}{=} \\ \mathsf{diagram}(\mathsf{square}, [1]) :: (4 \otimes \biguplus_{j=0}^{n+1} \mathsf{diagram}(\mathsf{row}, [2j]))\end{gathered}$$

□

9.2 Evaluation Issues

Evaluating the success of the ideas presented in this book and their implementation in DIAMOND can be done by evaluating the hypothesis for this work: DIAMOND *and the techniques developed in this book provide a feasible way of proving a limited class of theorems of mathematics.* The question now is whether the set of diagrams and operations available in DIAMOND enables one to prove diagrammatically a sufficient number of theorems. We discuss next the criteria for assessing that a number of proved theorems is sufficient.

9.2.1 Range, Depth and Source of Theorems

In Chapter 6 we described the operations in DIAMOND and claimed that the set of available operations should enable us to prove theorems of significant depth and range. The definitions of both significant depth and significant range are informal. By significant depth we hope to capture a set of examples which are not trivial to prove diagrammatically. This excludes theorems which require one-step non-recursive proofs, so the number of inference steps does not depend on the parameter. However, an exception to this rule are one-step non-recursive diagrammatic proofs (e.g., *commutativity of multiplication*) of theorems which are not trivial to prove with the usual symbolic logical machinery (e.g., due to the need for lemmas which may not be available, or the need for generalization). Theorems whose schematic proofs are non-recursive, but the proof consists of several inference steps, i.e., diagrammatic operations, are of significant depth. All theorems whose proofs are defined recursively, so the number of inference steps in the proof depends on the parameter for which the proof is given, are also of significant depth. Moreover, proofs of theorems which use other proofs as lemmas are of significant depth.

By significant range we mean to capture a variety of examples which are different from each other. For example, we claim that the set which contains recursively and non-recursively defined proofs is of significant range. Other criteria for the range include a variety of theorems about different natural numbers. For instance, proofs of theorems about square numbers, triangular numbers, Fibonacci numbers, hexagonal numbers, etc. form a set of proofs of significant range. Note that all of the mentioned proofs of theorems contribute not only to the range but also to the significant depth of proved theorems.

Our main source of examples are Nelsen's books *Proofs Without Words* (1993) and *Proofs Without Words II* (2001). We also found some examples in Penrose 1994a, Lakatos 1976, Gardner 1986, Gardner 1981, Dudeney 1958 and Gamow 1988. Our choice of theorems of natural number arithmetic rather than geometry means that the proofs which are considered to be *formal* proofs of these theorems are usually symbolic rather than diagrammatic proofs. Nelsen's books indicated a way to prove some of these theorems diagrammatically. Also, a theorem of geometry usually has an obvious diagrammatic representation, whereas a theorem of natural number arithmetic may not. Often, Nelsen's books helped us find a diagrammatic representation of our chosen theorems.

9.2.2 Methodology

The evaluation of DIAMOND consists of two stages. The first stage checks how many schematic proofs we can construct using DIAMOND. The second stage checks how many of these schematic proofs can be verified.

For the first stage we check whether DIAMOND is able to construct a schematic proof from example-proofs. This stage tests the expressiveness of the available diagrams and operations (see Chapter 5 and Chapter 6), and the capability of the abstraction mechanism (see Chapter 7).

The second stage checks whether the schematic proof is correct or not. The verification proof is carried out in the theory of diagrams (see Chapter 8). The verification of a schematic proof is done automatically.

Another test by which we can evaluate DIAMOND is to compare it with other theorem provers which construct proofs diagrammatically. However, not much work has been done on the automation of diagrammatic theorem provers. Some of the relevant work was surveyed in §2. An in-depth comparative analysis of systems related to DIAMOND, where it is evident that DIAMOND is not a rival to these other systems, can be found in Jamnik 1999.

9.2.3 Theorems Proved

We now list some of the theorems that we proved using DIAMOND. We distinguish between the development and the test set of theorems. This ensures that DIAMOND has not been specialized for only a few theorems during the development stage. If a number of theorems from the test set is successfully proved, or at least their schematic proofs can be found, then we can conclude that DIAMOND is reasonably general.

We first expresses a theorem in the usual symbolic way. In some cases, we represent a theorem with both the use of ellipsis and with the use of summation symbol $\sum$ together with the general form of a term. The ellipsis depicts the first few numbers, as well as the general form of a number in the sequence in the summation. $\sum$ captures ellipsis in an alternative way. Then, we state whether DIAMOND was capable of finding a schematic proof for the particular theorem under consideration. Next, we state whether the schematic proof which DIAMOND found is defined recursively or non-recursively. Finally, we state whether the schematic proof of a given theorem was successfully checked to be correct in the theory of diagrams.

Development Set of Theorems

1. $n^2 = 1 + 3 + 5 + \cdots + (2n-1) = \sum_{i=0}^{n}(2i-1)$: schematic proof was found and is recursive; its correctness was proved.
2. $\frac{n(n+1)}{2} = 1 + 2 + 3 + \cdots + n = \sum_{i=0}^{n} i$: schematic proof was found and is recursive; its correctness was proved.
3. $Tri(2n+1) = Tri(n+1) + 3Tri(n)$: schematic proof was found and is non-recursive; its correctness was proved.

Test Set of Theorems

1. $Tri(2n) = 3 + 7 + 11 + \cdots + (2(2n) - 1) = \sum_{i=0}^{n}(2(2i) - 1)$: schematic proof was found and is recursive; its correctness was proved.
2. $(2n + 1)^2 = 1 + (8 + 16 + \cdots + 4(2n)) = 1 + 4\sum_{i=0}^{n}(2i)$: schematic proof was found and is recursive; its correctness was proved.
3. $Fib(n) \times Fib(n + 1) = Fib(1)^2 + Fib(2)^2 + \cdots + Fib(n)^2 = \sum_{i=1}^{n} Fib(i)^2$: schematic proof was found and is recursive; its correctness was proved.
4. $Tri(2n-1) = 1+5+9+\cdots+(2(2n-1)-1) = \sum_{i=0}^{n}(2(2i-1)-1)$: schematic proof was found and is recursive; its correctness was not proved.
5. $2n - 1 = (\sum_{i=1}^{n-1} 2) + 1$: schematic proof was found and is recursive; its correctness was not proved.
6. $n(n + 1) = \frac{n(n+1)}{2} + \frac{n(n+1)}{2}$: schematic proof was found and is non-recursive; its correctness was proved.
7. $Tri(2n) = Tri(n - 1) + 3Tri(n)$: schematic proof was found and is non-recursive; its correctness was proved.
8. $(2n + 1)^2 = 8Tri(n) + 1$: schematic proof was found and is non-recursive; its correctness was proved.
9. $(2n)^2 = 8Tri(n - 1) + 4n$: schematic proof was found and is non-recursive; its correctness was proved.
10. $n \times (n + 3) = (n + 3) \times n$: schematic proof was found and is non-recursive; its correctness was proved.

Note, that the theorem $n(n + 1) = \frac{n(n+1)}{2} + \frac{n(n+1)}{2}$ is an instance of a theorem $x = \frac{x}{2} + \frac{x}{2}$. The former theorem is about two triangles of equal magnitudes n which together form a rectangle of magnitude n by $n + 1$. The latter theorem is from a diagrammatic point of view more general, where $x = n(n + 1)$.

Furthermore, notice that theorem $n \times (n + 3) = (n + 3) \times n$ is an instance of a theorem $n \times m = m \times n$ which is universally quantified over two parameters and so it cannot be proved in DIAMOND. We arbitrarily chose to instantiate m to $n + 3$, but any other instance of m would have the same diagrammatic proof.

DIAMOND used five different kinds of diagrams and eleven diagrammatic operations to prove the theorems listed above. All together, we proved around thirty-five theorems of significant range and depth with DIAMOND. All of the diagrammatic proofs for theorems listed here are interesting in that they are easily intuitively understood. Furthermore, their diagrammatic proofs have not been mechanized before.

9.2.4 Diamond's Limitations

The number of theorems that Diamond to date can prove is limited by various factors. These include limitations of:

- the number of diagrams and operations available to users,
- the abstraction mechanism, since Diamond's abstraction cannot deal with more than one variable,
- the verification module due to weaknesses in the implementation of the theory and due to weaknesses in *Clam*,
- the user interface, since Diamond cannot display diagrams in three dimensions.

We discuss each of these in turn.

Limitations on the Diagrams and Operations

There are about eight different diagrams and fourteen different operations available in Diamond. As discussed in §9.2.1, these should enable us to prove theorems of significant range and depth, and as showed in §9.2.3 they do indeed. Clearly, implementing additional diagrams and operations would allow us to prove more theorems. For example, if we had an operation available which splits a triangle into isosceles trapeziums, as in Figure 41 then we would be able to prove, e.g., at

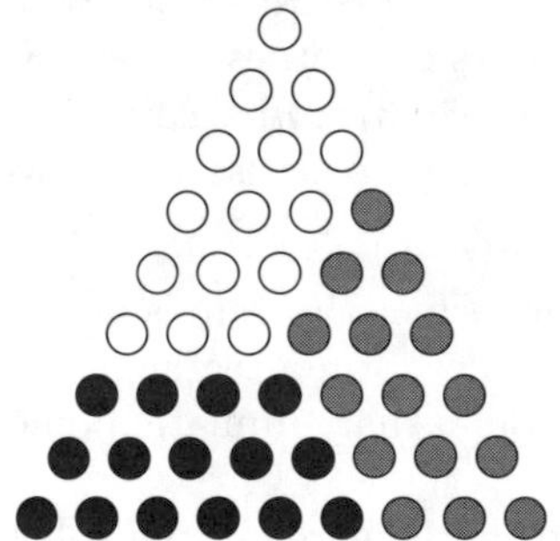

FIGURE 41 $Tri(3k) = 3(Tri(2k) - Tri(k))$

least these theorems: $Tri(3k) = 3(Tri(2k) - Tri(k))$ and $Tri(3k+2) = 3(Tri(2k+1) - Tri(k))$, which are taken from Nelsen 2001. Furthermore, if we had diagrams such as pentagons available in Diamond, then we would be able to prove theorems about pentagonal numbers, etc.

The question is whether such additions contribute to the range and depth of diagrammatic proofs we can construct in Diamond. Also, additional operations must be generally useful and not *ad hoc*. It is our heuristic choice to limit the set of diagrams and operations to the ones which are implemented in DIAMOND to date. Potentially, additional or new ones which subsume the existing ones can be added to the set, but this remains one of the tasks for future work. The justification for such a heuristic choice is that with currently available diagrams and operations we are able to prove theorems of significant range and depth, as showed in §9.2.3.

Limitations of Abstraction Mechanism

DIAMOND's abstraction mechanism has several weaknesses which limit the kind of schematic proofs which it is capable of constructing. We mention here only one such limitation , namely the inability to construct a schematic proof due to a more complex structure of a schematic proof than the one formalized in DIAMOND.

Let us recall DIAMOND's formalization of a schematic proof which was given in equations (7.4) and (7.5).

$$\begin{aligned} \mathsf{proof}(n+1) &= \mathcal{A}(n+1),\ \mathsf{proof}(n) \\ \mathsf{proof}(0) &= \mathcal{B} \end{aligned}$$

Now, consider a theorem which is stated as $(2^n)^2 = \Pi_{i=1}^{n}(4 \times 1^2)$, and its example diagrammatic proof for $n = 3$ given in Figure 42. The diagram-

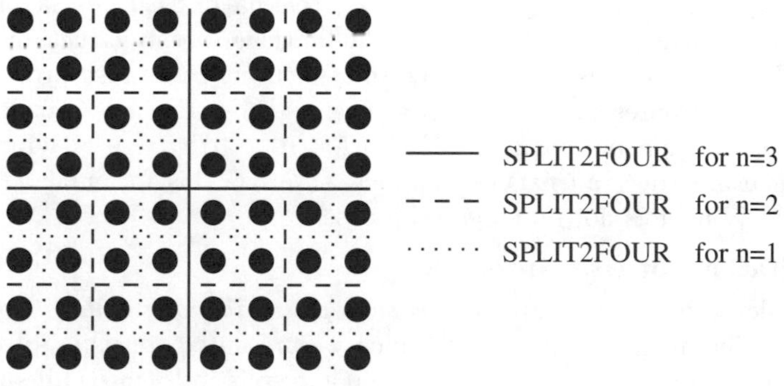

FIGURE 42 $(2^n)^2 = \Pi_{i=1}^{n}(4 \times 1^2)$

matic proof consists of applying the operation split2four an appropriate number of times. The schematic proof of this theorem can be formalized as:

$$\begin{aligned} \mathsf{proof}(n+1) &= [(\mathsf{split2four}, 1)],\ \mathsf{proof}(n),\ \mathsf{proof}(n),\ \mathsf{proof}(n),\ \mathsf{proof}(n) \\ \mathsf{proof}(0) &= [\,] \end{aligned}$$

DIAMOND's abstraction mechanism is not powerful enough to be able to construct this type of complicated formalization of a schematic proof. Any other structure of a schematic proof than the one given in equations (7.4) and (7.5) cannot be constructed by DIAMOND's abstraction mechanism. This limits the range of theorems that DIAMOND is capable of proving. The limitation can be removed by extending the possible formalizations of schematic proofs and hence, extending the abstraction mechanism accordingly.

Limitations of Verification Mechanism

It is evident from the list of proved theorems in §9.2.3 that not all schematic proofs could be verified automatically. We carried out an experiment on paper, and used the theory of diagrams discussed in Chapter 8 to verify whether the schematic proofs that were not proved automatically to be correct are indeed correct. We found that all of the schematic proofs were correct.

One of the reasons that the verification of all schematic proofs cannot be carried out automatically is due to the limitations of the implementation of the theory in *Clam*, and due to the limitations of *Clam* itself. Namely, *Clam* is not very good with arithmetic rewriting and non-zero conditionals in the induction strategy. For instance, *Clam* finds it difficult to find a proof plan for the theorem which is not quantified over all natural numbers but only over non-negative naturals. *Clam* is also not very good in using non-constructive definitions in the induction strategy. For example, using a predecessor functions can cause problems – rather than representing $Tri(2n) = Tri(n-1) + 3Tri(n)$ *Clam* prefers a representation of $Tri(2(n+1)) = Tri(n) + 3Tri(n+1)$ which can be quantified over all natural numbers. We use this formulation when possible, but it is not possible in all cases.

Limitations of User Interface

Consider the theorem about the *sum of hexagonal numbers* given in §3.2.5. The diagrammatic proof which we presented consisted of taking a cube, looking down the main diagonal and splitting it into half-shells,[60] and finally for each half-shell, we project it from three dimensions onto a plane and observe that it forms a hexagon.

To be able to construct this diagrammatic proof, we need to have a three dimensional environment in which diagrams such as cubes, and operations such as splitting a half-shell from a cube are available to us. To date, DIAMOND's interface is capable of displaying two-dimensional images only. This clearly limits the number of theorems which can be

[60]Recall that a half-shell consists of three adjacent sides of a cube.

proved by DIAMOND. However, we proposed to Farrow (1997) to design a three-dimensional (3D) diagrammatic viewer which is capable of displaying two and three dimensional diagrams and operations on them. Our idea was to link such a viewer to DIAMOND, so that the example-proofs are constructed in three dimensions using the viewer, but the schematic proof is abstracted and verified in DIAMOND. Figure 43 shows

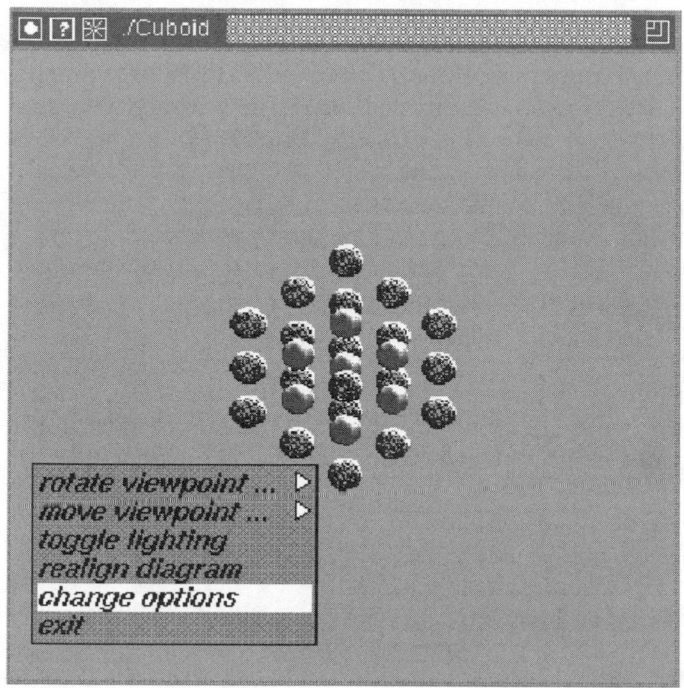

FIGURE 43 An example of three-dimensional virtual environment for diagrammatic proofs.

how a cube is observed down its main diagonal after being split into half-shells. This viewing makes the fact that a half-shell forms a hexagonal number explicit.

The 3D diagrammatic viewer allows a user to construct example-proofs in a similar way as DIAMOND. In the end, it produces example-proof traces which can be passed to DIAMOND, so that DIAMOND's abstraction mechanism can attempt to construct a schematic proof. Here is an example-proof of a theorem about *sum of hexagonal numbers* that the 3D diagrammatic viewer allows the user to construct (note that the proof trace is given for $n = 4$):

$$\begin{aligned}\mathsf{proof}(4) = [&(\mathsf{split\_half\_shell}, 1), (\mathsf{project\_to\_2d}, 1),\\ &(\mathsf{split\_half\_shell}, 1), (\mathsf{project\_to\_2d}, 1),\\ &(\mathsf{split\_half\_shell}, 1), (\mathsf{project\_to\_2d}, 1),\\ &(\mathsf{split\_half\_shell}, 1), (\mathsf{project\_to\_2d}, 1)]\end{aligned}$$

The proof trace for $n = 3$ is the same as this minus the first two operations:

$$\begin{aligned}\mathsf{proof}(3) = [&(\mathsf{split\_half\_shell}, 1), (\mathsf{project\_to\_2d}, 1),\\ &(\mathsf{split\_half\_shell}, 1), (\mathsf{project\_to\_2d}, 1),\\ &(\mathsf{split\_half\_shell}, 1), (\mathsf{project\_to\_2d}, 1)]\end{aligned}$$

Using the abstraction algorithm given in §7.5, the following schematic proof can be constructed (recall that it has not been implemented yet):

$$\begin{aligned}\mathsf{proof}(n+1) &= [(\mathsf{split\_half\_shell}, 1), (\mathsf{project\_to\_2d}, 1)],\ \mathsf{proof}(n)\\ \mathsf{proof}(0) &= [\,]\end{aligned}$$

Finally, additional definitions for new diagrams and operations need to be added to the theory of diagrams in order to verify this schematic proof. We foresee no particular obstacles in implementing these additions. Hence, in principle, using the formalization of proofs used in DIAMOND, we can abstract a diagrammatic proof of a theorem about the *sum of hexagonal numbers* as presented by Penrose (1994a).

9.3 Summary

In this chapter we presented DIAMOND in action in order to demonstrate some of the results we achieved by implementing it. The success of our approach can be evaluated by determining the depth and the range of theorems that DIAMOND can prove. In general, we can conclude that DIAMOND is successful in diagrammatically proving a significant range and depth of theorems.

There are many reasons that prevent DIAMOND from proving more theorems of natural number arithmetic. These include the insufficient number of available diagrams and operations, the limitations in the abstraction mechanism, the weaknesses in the implementation of the verification mechanism, and the limitations of the user interface. Improving them would enable DIAMOND to prove more theorems.

10

Complete Automation

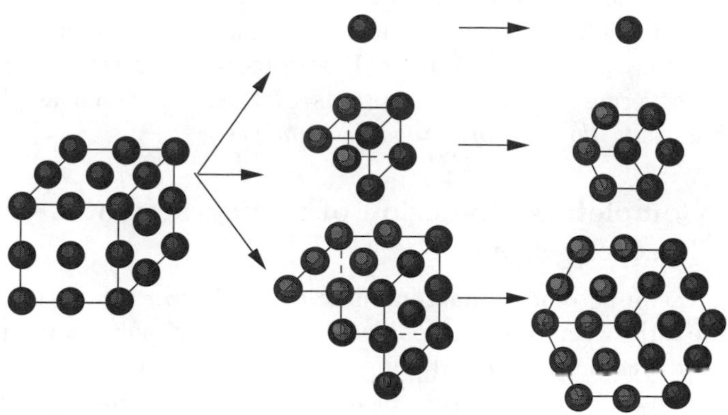

$$n^3 = 1 + 7 + 19 + \cdots + Hex(n)$$

— ANON
in PENROSE's *Mathematical Intelligence*
and in NELSEN's *Proofs Without Words*

We hope that we have convinced the reader that diagrams are a reasoning tool worthy of further investigation. Despite the fact that humans frequently use them when proving theorems, diagrams are not generally accepted as a tool for formal reasoning. In an effort to change this, we have shown that diagrams can be used in formal proofs, and moreover that diagrammatic reasoning can indeed be formalized and emulated on machines. Our formalization of diagrammatic reasoning is embodied in a semi-automatic system DIAMOND which allows a user to construct diagrammatic proofs of mathematical theorems. Our work is also a response to Penrose's challenge in which he claimed impossible to automate diagrammatic proofs, such as the one he presented regarding the *sum of hexagonal numbers* (demonstrated in the picture above). DIAMOND cannot automatically find a diagrammatic proof of this particular theorem yet, because the interface which allows manipulations of diagrams in three dimensions (see §9.2.4) has not been added to it. Nevertheless, we have shown that the *approach* to reasoning embodied in Diamond allows automation of a theorem prover which proves theorems, including the one given by Penrose, diagrammatically.

Of course, Penrose may object further that Diamond cannot discover diagrammatic proofs, and that most of the intelligence in Diamond is provided by the user, since it is the user who constructs the example-proofs. However, the work presented in this book makes first steps on the way to explore the nature of human "informal" mathematical reasoning. These steps shed light on how a system *could discover* such proofs, and furthermore, they demonstrate that this is indeed *possible*. In this chapter we describe how Diamond could be extended from a semi-automatic proof checking system to a completely automated theorem prover capable of discovering diagrammatic proofs of theorems, such as the one above regarding *the sum of hexagonal numbers*.

10.1 Complete Automation of a Diagrammatic Theorem Prover

Making a semi-automatic diagrammatic proof checker into a completely automated theorem prover capable of discovering new diagrammatic proofs is an ambitious undertaking, but one that takes the work described here to a higher level. Such a project will contribute to two parallel research aims of artificial intelligence and computer science: on the one hand it will create new technology that will enable computers to achieve new goals and solve new problems, and on the other hand it will help us to better understand diagrammatic reasoning.

We describe only the general methodology which could be employed in order to achieve complete automation of a diagrammatic theorem prover. The intricacies of the system are left for investigation in the future.

A starting point for the proposed research will be the same as in the development of Diamond: we will need to choose a problem domain with a wealth of theorems that can be represented and proved diagrammatically. This could be the same as Diamond's, and any undergraduate mathematical text, especially in the domain of natural number arithmetic would be a good source of problems. Next, a set of possible diagrams and diagrammatic operations will need to be identified. As in Diamond, we should define a mapping relation between these diagrams and symbolic formulae, and these operations and formulae. To begin with, only basic operations will be required, such as combining rows and columns. More complex operations could be at first added by the developer of the system, or may later be discovered by the system automatically.

Having appropriate diagrammatic operations available to the system is important for the system's ability to prove theorems. If a required operation is not available, then the system may not be able to find a proof of the theorem at hand. The developer of the system needs to add this operation to the system or make it explicitly available to the user who constructs the proof. With the proposed system, some operations may be automatically discovered as compositions of other existing operations, and then added to the set of all available operations for use in the discovery of new proofs.

The next step in formalizing the proposed reasoning framework will be to construct examples of diagrammatic proofs from which new proof methods will be learned. The conjectures could be input either as symbolically represented theorems by the user, or interesting combinations of diagrams can be constructed automatically by the system. In the case of the symbolic input conjecture, a mapping relation, much the same as DIAMOND's, could be used to map the symbolically represented theorems into diagrammatic representation. Clearly, in the case of an automatic construction of a conjecture, we will need to define some criteria as to what is "interesting".

Next, the available diagrammatic operations will be applied in order to construct diagrammatic proofs. Again, these could be input by the user, or constructed automatically. In the case of automatic construction, the operations will be applied so long as an "interesting" proof is found. Initially, the user could decide whether the proof of a theorem is of interest or not. Later, some criteria can be implemented to decide the interestingness of new proofs. For example, are the operations applied recursively, or are some newly constructed operations used, or are existing operations used in a new way, or are the sequences of diagrammatic operations long? An affirmative answer to any of these questions may indicate that the newly constructed proof is interesting.

The construction of new diagrams and operations is closely related to the construction and the discovery of new proofs and indeed theorems too. Namely, every new diagram and hence new operation on it (recall that we demonstrated in Chapter 6 the close relation between diagram representations and operations on diagrams) could give us clues as to what symbolic expression it represents. Furthermore, it could also be seen as a new theorem and hence also its proof. For example, if we did not have a square frame diagram available is DIAMOND (see 6.2), then such a frame could be constructed using combinations of joins between columns and rows. In relation to the magnitude of the frame, say it is n, the four columns and rows are of magnitude $n - 1$. Then, a frame could in general represent the symbolic expression $4(n - 1)$. However, this frame could also be constructed by removing an inner square from a given square of magnitude n. Hence, the frame could also represent the symbolic expression $n^2 - (n - 2)^2$. Thus, by discovering a different construction of a diagram, we also discovered a new theorem, namely that $4(n-1) = n^2 - (n-2)^2$. Moreover, we also demonstrated that this theorem holds.

The user or the machine could discover such new diagrams and operations, and hence new theorems and their diagrammatic proofs, by exploring different combinations of existing diagrams and operations. It is not necessary that this exploration is goal directed, i.e., that the user or the machine has a specific theorem in mind in order to search for a new proof. Rather, the system could allow a systematic exploration of the structure of diagrams in the hope that according to some criteria for interestingness mentioned above, new theorems and their proofs would emerge. This exploration is described by Lindsay (2000) as the so-called *play* with diagrams.

The automatic exploration and construction of example-proofs could be administered, to start with, by an exhaustive search amongst all the available diagrammatic inference steps. Given simple initial diagrams and a limited repertoire of geometric operations, this may be tractable – our work indicates that diagrammatic reasoning in some domains, e.g., a particular subset of natural number arithmetic used in this book, may be more effective than symbolic reasoning.

Many have argued about the efficacy of diagrammatic reasoning (Lemon et al. 1999). It seems that diagrammatic representations are *sometimes* more effective and efficient than symbolic formulae. This may be due to the fact that they store information more concisely, and they explicitly represent the relations amongst the elements in the diagram (Larkin and Simon 1987). For example, complex symbolic manipulations may be required in order to symbolically deduce that $n^3 = (n-1)^3 + Hex(n)$. However, when using diagrams this can be established simply by removing a half-shell from a cube. Furthermore, n^2 does not have alternative symbolic representations, whereas a diagram which is a square can implicitly denote other equivalent symbolic representations as well, e.g., $n^2 = n \times n = Tri(n) + Tri(n-1) = \cdots$. These can readily be used in reasoning and do not need to be computed explicitly, as in the case of symbolic reasoning. Moreover, diagrams and operations on them can be viewed as macro inference steps. For example, if a square is represented as a nested sequence of ells, then its decomposition into ells can readily be achieved. On the other hand, in symbolic reasoning, inference steps need to be applied explicitly one at a time. There are, of course, domains for which symbolic reasoning is more efficient than diagrammatic reasoning. These domains typically deal with expressions which cannot be easily represented as diagrams, e.g., let us just try to imagine diagrams in four or more dimensions.

An exhaustive search for new theorems and proofs, which was suggested above as the first step to complete automation, could later be improved and controlled by the use of heuristics. For instance, certain known geometric properties could indicate how to prune the search. Here are some examples:

- A heuristic which determines which particular kind (i.e., version) of an operation is possible or is sensible to be applied to the diagram (e.g., it is sensible to cut a square so as to preserve right angles).
- A heuristic which detects when it is impossible to generate a certain construction with the available operations on a particular diagram (e.g., it is impossible to split a discrete square of odd magnitude into four identical squares).
- Additional pieces of knowledge of certain geometric properties of diagrams (e.g., a pentagon cannot be split into five equilateral triangles).
- A heuristic which prefers compositions and decompositions which lead to preserving the recursive structures of diagrams (e.g., it is sensible to respect the recursive definition of a square whenever possible).
- A heuristic which prefers new compositions and decompositions of *existing* diagrams, rather than arbitrary new ones.
- A heuristic which may identify certain frequently used compositions of diagrams and hence focus the search for the proof on this particular part of the diagram (e.g., ideas from the DC model, see §2.2, could be used here).

The use of these heuristics in proof search could be guided by the existing techniques, such as proof planning techniques (Bundy 1988). Using the heuristics, several possible sequences of operations on concrete diagrams could be generated automatically. Subsequently, they will need to be abstracted to form a general proof. It seems plausible that an appropriate representation of the problem might give us clues as to how to achieve this. George Pólya in his book *Mathematical Discovery* (1965) argues that the choice of the representation of the problem is vital to finding its solution. We would like to consider this suggestion more closely. It appears that different representations of the same diagram will lead to a discovery of different diagrammatic proofs, and subsequently different formal mathematical proofs.

If our work is extended as suggested in this section, and a completely automated diagrammatic theorem prover is implemented, then we think that such a system would be able to discover diagrammatic proofs of some mathematical theorems. Such a system would be a direct response to Penrose's challenge (see §4.6). Moreover, the hope is that it will lead us to the discovery of not only proofs of known theorems, but also to the discovery of new and interesting theorems and their proofs. In this process, the concept of interestingness of a theorem or proof will need to be explored and defined. This will shed light on the question of what is it in diagrammatic reasoning that is intuitive and easily understandable to humans. Ultimately, it may provide us with further clues about the cognitive model of human reasoning.

10.2 The Context of Our Work

The aims of our work can be placed in the context of several research directions. We implemented a semi-automated *reasoning system* which reasons with geometric manipulations of diagrams. When complete diagrammatic proofs are constructed in DIAMOND, we have a guarantee that they are correct. Hence, we showed that diagrams can be used for *formal proofs*. Finally, these formal proofs are constructed in DIAMOND from *examples of proofs*, hence the concreteness property of diagrams can be readily exploited. We discuss each of these issues next.

10.2.1 Automating Diagrammatic Reasoning

The first automated reasoning systems were implemented in the fifties when Newell and Simon built a program that could prove simple theorems of propositional logic (Newell and Simon 1956). There has been a lot of interest since in the automation of theorem proving, and as a result we nowadays have very many complex systems including Nqthm by Boyer and Moore (1990), Isabelle by Paulson (1989), Nuprl by Constable et al. (1986), and *Clam* by Bundy et al. (1991). All of these reasoning systems use the usual symbolic *logical* representations, such as sequent calculus, for mathematical reasoning. The systems use the rules of some chosen logic in order to generate a proof of a theorem of mathematics.

In subsequent years, formal mathematical logic has been considered as one of the very few tools which is rigorous enough to base automated reasoning systems on. A more "informal" aspect of human mathematical reasoning, such as the use of diagrams to convey truths of statements, has been neglected. However, in the past two decades, researchers have looked into how more "informal" aspects of human mathematical reasoning, especially the use of diagrams, can be incorporated into automated reasoning systems. In particular, one of the first systems to use diagrams to guide a search for proofs was Gelernter's Geometry Machine (1963). The systems which have been devised since (e.g., Grover by Barker-Plummer and Bailin (1992), and Hyperproof by Barwise and Etchemendy (1994)) use diagrams to model the problem and to guide the search for what is essentially a symbolic logical proof. There is also some work which shows that diagrams can be used for formal reasoning by devising a logic of diagram (e.g., Shin 1995 and Hammer 1995).

Our research and the implementation of DIAMOND is new in the area of automated reasoning, since it uses only diagrammatic inference rules in the construction of proofs, and these proofs are guaranteed to be correct. The usual symbolic logical inference rules are replaced in DIAMOND by geometric operations on diagrams. Rather than constructing proofs by chaining together logical formulae, proofs in DIAMOND are constructed by applying various combinations of geometric operations to diagrams. Unlike most existing theorem provers which use only logical formulae in proofs, DIAMOND uses only geometric operations to construct proofs. Moreover, unlike the Geometry Machine and other systems that use a combination of symbolic and diagrammatic inference rules, DIAMOND uses only diagrammatic inference rules. No logical formulae are needed when constructing proofs. Finally, unlike in most of the systems that *do* use diagrammatic inference steps (e.g., Archimedes by Lindsay (1998), Bitpict by Furnas (1992), and IDR by Anderson and McCartney (1995)), the construction of proofs in DIAMOND is supported by machinery which ensures that DIAMOND's diagrammatic proof is a correct proof of a theorem.

In the implementation of DIAMOND we addressed the general issues of diagrammatic versus symbolic representations and reasoning. In DIAMOND multiple representations of diagrams have been devised. Their use is equivalent to using different representations of a problem. The solution to a problem can be obtained depending on whether the right representation is used. Such an approach to knowledge representation illuminates how to use Pólya's advice on the importance of appropriate representations (Pólya 1965). We characterize diagrammatic and symbolic representations in so far as it allows us to characterize the differences between reasoning with chains of formulae and reasoning with diagrams. But there may be other types of representations as well, e.g., linguistic or audible. Furthermore, there is no line that draws a clear distinction between diagrammatic and symbolic reasoning. Rather, there are varying degrees to which a representation is symbolic or diagrammatic. Indeed, some would argue that writing sentences or computing multiplications of numbers is to some degree diagrammatic since the letters and the numbers need to be spatially related to one another. In this book, rather than drawing distinctions, we illuminated the characteristics of various types of representations and reasoning. In particular, we demonstrated how diagrams allow us to change the representation of a problem to make the task of finding its solution easy and intuitive. Sometimes a combination of diagrammatic and symbolic reasoning may be more intuitive than just one or the other alone. I this book we identified an area for which diagrammatic reasoning is at times more understandable than symbolic reasoning, because it fits both, the perceptual processes humans are good at, and our cognitive knowledge about spatial structures that we experience around us.

10.2.2 Can Diagrams Be Used In Formal Proofs?

Diagrams have been used as an aid in reasoning for centuries. At the turn of this century the invention of rigorous axiomatic logical reasoning made a significant impact on the notion of formal reasoning. Part of this influence was a belief that diagrams are not rigorous enough to be used as a tool in formal reasoning. However, in the last two decades this belief has changed, and we can observe an increased interest in research on re-establishing a formal role for diagrams in reasoning.

Our semi-automatic proof system Diamond is a realization of a formalization of diagrammatic reasoning. Our research contributes to the effort of showing that diagrams can indeed be used as a tool for *formal* automated rather than just informal human mathematical reasoning. Diamond provides an environment in which diagrammatic proofs of mathematical theorems can be constructed. The method of diagrammatic proof construction in Diamond consists of three steps:

1. The user can construct instances of a diagrammatic proof using various combinations of diagrams and operations applied to them. All diagrams are concrete, drawn for a particular value of a universally quantified variable.

2. Diamond automatically constructs a general diagrammatic proof from these instances. Diamond's diagrammatic proof is captured by a recursively defined schematic proof, and consists of a general number of applications of geometric operations.

3. In Diamond we have machinery which can formally show whether a diagrammatic proof is correct or not. This machinery is embodied in a theory of diagrams in which Diamond can automatically formally verify a constructed diagrammatic schematic proof.

The construction of a schematic proof is an educated guess made by a machine of what looks like the most likely proof of a theorem, given some example-proofs. The diagrammatic theory which is provided in DIAMOND is a formal theory in which DIAMOND can check that this guess was indeed correct. In this way, we ensure that a diagrammatic proof is a correct proof of a theorem in a formal logical sense.

One of the conclusions that we can draw from our work is that the neglect of the use of diagrams in reasoning due to the belief that diagrams are not a formal or rigorous enough device is not justified. By implementing DIAMOND we show that diagrammatic reasoning can be formal.

10.2.3 Diagrammatic Proofs

The research in machine learning techniques for generalization or in our terminology, abstraction, was a vibrant area in the sixties and seventies. One of the first algorithms for abstraction was devised by Plotkin (1969, 1971) which attempts to find the least general term from specific examples. Since then, many variations of Plotkin's abstraction algorithm have been invented, all specialized for particular classes of problems.

The abstraction mechanism in DIAMOND is a variation of Plotkin's and Baker's (1993) algorithms. It extends them to allow the abstraction of functions of applications of inference rules and the recursive structure of the schematic proof, and it can do so by using only two examples.

Baker explored the use of the constructive ω-rule, a stronger alternative to the induction rule, for logical proofs of arithmetic theorems. The constructive ω-rule requires the provision of a uniform computable procedure which by instantiation produces a proof for a corresponding instance of a theorem. Baker used schematic proofs to provide this uniform procedure. The use of the constructive ω-rule provides a technique for constructing proofs from instances of proofs by providing a justification that a correct schematic proof is a formal proof of a theorem.

We extended Baker's work from arithmetic theorems to diagrammatic reasoning. Using the constructive ω-rule in schematic proofs allows reasoning with instances of a diagrammatic proof. Therefore, the diagrams which are employed in instances of a proof can be concrete rather than abstract. In this way we avoid the need for a formalization of abstraction devices (e.g., ellipsis) in general diagrams, and so avoid difficulties with such representations.

Rather than using meta-induction to verify schematic proofs, as Baker did, we devised a diagrammatic theory where schematic proofs can be checked for their correctness without any need for meta-induction. Meta-induction on diagrams is open to problems because it requires reasoning with general diagrams which use abstraction devices. In our theory of diagrams, the verification of schematic proofs seems to require only simple standard mathematical induction, while at the same time it removes the need to formalize abstraction devices in diagrams.

This framework for constructing diagrammatic proofs has an advantage over other existing diagrammatic reasoning systems (see the survey in Chapter 2). Namely, if all goes well in the construction of a diagrammatic proof, then the proof in the end is guaranteed to be correct.

10.2.4 The Human Mathematician and DIAMOND

The implementation of DIAMOND is a valuable research project in its own right which is evident by the contributions made to several aspects of computer science mentioned in the preceding few sections. Here, we propose to cognitive scientists the potential for using a DIAMOND-like system in experiments which would test the psychological validity of diagrammatic reasoning. The implications of a potentially positive result of such testing could have an impact on various fields, but especially in teaching students mathematics.

One of the aims of our work is to model "informal" human reasoning processes, in particular, the use of diagrams in proving mathematical theorems. This can be approached from the cognitive or computational directions, as we explained in the introduction to this book. From the cognitive perspective, different cognitive models of human reasoning can be investigated. These cognitive models can then serve to inform the computational models how to make machines think more like humans. From the computational perspective, we can design computational frameworks which model human cognitive processes. They can be a source for empirical testing that shows how consistent is the computational model with the human cognitive processes. In this book we took the latter approach.

DIAMOND provides an architecture for the construction of diagrammatic proofs. Our belief is that diagrammatic proofs of the kind that we presented in this book are more easily understood by humans than their corresponding symbolic proofs. However, we have not carried out any psychological validity testing on human mathematicians which would empirically support our belief. DIAMOND provides an architecture which could be used by cognitive scientists as a basis for testing to what extent novice or expert mathematicians find diagrammatic proofs more intuitive than symbolic proofs. Moreover, DIAMOND can be used as an environment which enables users to explore reasoning with diagrams and thus, hopefully gain an understanding into a diagrammatic proof of a theorem.

If the outcome of such a comparative study supports our belief, this could have educational implications. Namely, it would draw attention to the types of cognitive mechanisms that students of mathematics need in order to reason rigorously with diagrams. A grasp of the nature of such mechanisms might help mathematics teachers develop new ways of engaging these mechanisms, e.g., developing mathematical intuitions. A DIAMOND-like system could be a useful teaching tool for doing so.

10.3 Have We Achieved the Aims?

We hope the reader is convinced by now that the aims which we set ourselves at the beginning of this book and were outlined in Chapter 1 have been achieved. We automated parts of mathematical reasoning with diagrams in the domain of natural number arithmetic — the implementation of our research is a semi-automatic proof system DIAMOND. Diagrammatic proofs are formed from examples that the user constructs by applying geometric operations to diagrams. DIAMOND extracts the structure common to these examples and represents it in the form of a recursive program, called a schematic proof, which consists of a general number of applications of operations to diagrams. We devised a theory of diagrams in which DIAMOND formally verifies the correctness of a schematic proof. If the schematic proof is correct, then it constitutes a formal diagrammatic proof of a theorem.

Our work shows that despite the fact that diagrams have been denied a formal role in theorem proving, they can be used as a formal tool for rigorous mathematical reasoning. Unlike many other existing systems, diagrams in DIAMOND are not used as a model of a problem to find an essentially symbolic proof of a theorem, but are used directly to reason *with* them.

There are many aspects of our research which could be developed and extended further. We have already started to look how the approach in DIAMOND could be extended to continuous space, and thus geometry and real analysis (Winterstein et al. 2000). Ultimately, we hope that DIAMOND could be extended to a theorem prover capable of discovering diagrammatic proofs for itself.

Appendix A: More Examples of Diagrammatic Theorems

Here we give more examples of theorems and their diagrammatic proofs. These examples, in addition to the ones given in Chapter 3 plus others from Nelsen 1993, Nelsen 2001, Lakatos 1976 and Gamow 1988, are analyzed to motivate a taxonomy of diagrammatic theorems for choosing a problem domain (see §3.5) considered in this book.

The examples are given in terms of diagrams to which operations are applied, followed by a description of the proof. Finally, we examine these proofs to identify the required repertoire of geometric operations, and to formalize the general structure of diagrammatic proofs. The following examples of theorems are presented: *Pythagoras' theorem*, two different *triangular equalities, sum of all naturals*, and *Euler's theorem*. Note that the proof of *Euler's theorem* is an example of an erroneous schematic proof. Some discussion about this proof is carried out in §4.5, but for more information, the reader is referred to Lakatos 1976. The theorems that a user can prove using DIAMOND are both *triangular equalities* and *the sum of all naturals*.

A.1 Pythagoras' Theorem II

Pythagoras' Theorem states that a square on the hypotenuse of a right-angle triangle equals the sum of the squares on its other two sides. Figure 44 shows a diagrammatic proof of this theorem which is different to the one given in §3.2.2, and is taken from Nelsen 1993, page 4.

$$c^2 = a^2 + b^2$$

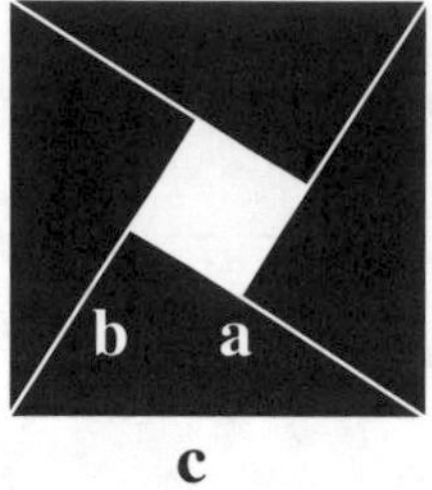

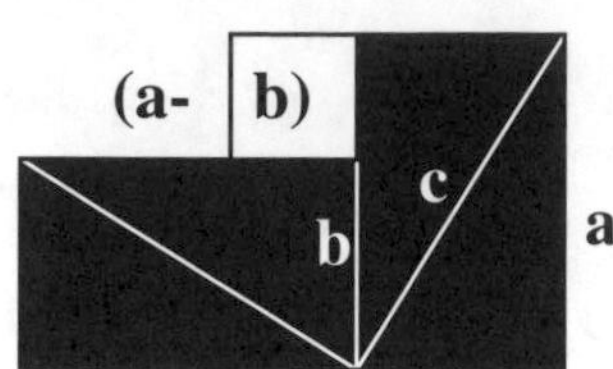

FIGURE 44 Pythagoras' theorem.

The diagrammatic proof demonstrated in the picture consists of taking any right angle triangle with a hypotenuse c and two sides a and b. We now join to this triangle along its shorter side another identical right angle triangle with its longer side touching the shorter side of the first triangle. We take a third identical right angle triangle and join to the second triangle in the same way as before. We repeat the same process for the fourth triangle which is joined to the third and also the first triangle. Notice that this process forms a square of magnitude c which is the hypotenuse. Note also that in the middle, there is another smaller square formed. Therefore, this formation justifies the following equation $c^2 = 4\frac{ab}{2} + small\_square$.

Now, we rearrange the diagram forming the square of magnitude c by moving two of those right angle triangles, and joining them along their hypotenuse to the remaining two triangles. Notice now that the smaller square is joined to the shorter side on a triangle to form its longer side, thus the magnitude of the smaller square is $a - b$. Hence we have $c^2 = 4\frac{ab}{2} + (a-b)^2 = 2ab + a^2 - 2ab + b^2 = a^2 + b^2$.

A.2 Triangular Equality for Odd Squares

The following is a proof of the *equality of triangular numbers for odd squares*. The example is taken from Nelsen 1993, page 101. The theorem and its diagrammatic proof are demonstrated in Figure 45.

$$(2n + 1)^2 = 8Tri(n) + 1$$

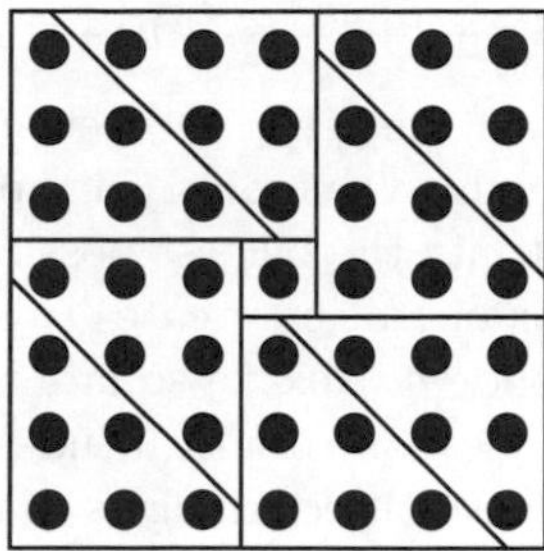

FIGURE 45 Triangular equality of odd squares.

The proof consists of taking a square of magnitude $2n + 1$ for a particular value of n. We then split the middle dot from it. This results in a thick frame. We split this frame into four rectangles. Note that two of the rectangles will be of magnitude $(n + 1) \times n$, and two of them will be of magnitude $n \times (n + 1)$. We split now each of the rectangles diagonally. This results in the formation of eight triangles of magnitude n. Considering the dot in the beginning we have $(2n+1)^2 = 8T(n)+1$.

A.3 Even Triangular Sum

The following is a proof of the *equality of even triangular numbers*. The example is taken from Nelsen 1993, page 104. The theorem and its diagrammatic proof are demonstrated Figure 46.

$$Tri(2n) = 3Tri(n) + Tri(n - 1)$$

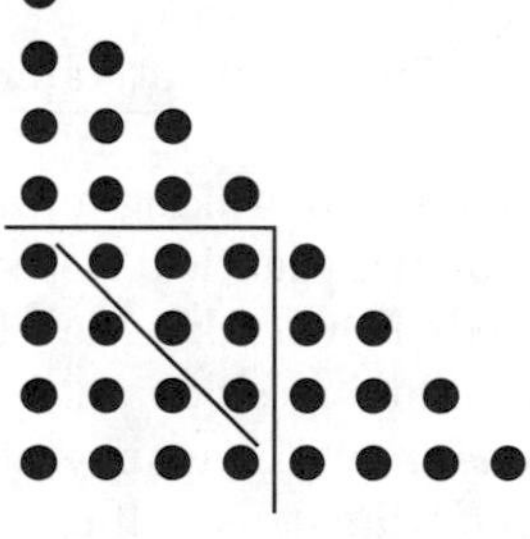

FIGURE 46 Even triangular sum.

Note that without using the definition of triangular number, this theorem could be restated into the following:

$$1+2+3+\cdots+2n = 3(1+2+3+\cdots+n)+(1+2+3+\cdots+(n-1))$$

The diagrammatic proof of the given theorem takes a triangle of magnitude $2n$ for some particular value of n (in the example above $n = 4$). This triangle is then split into the biggest possible square and two other triangles. Notice that this operation creates two triangles of magnitude n and a square of magnitude n. Next, a square is split down the middle, which results in two triangles, one of magnitude n and the other of magnitude $n-1$. Hence, we have three triangles of magnitude n and one of magnitude $n-1$.

A.4 Sum of All Natural Numbers

The theorem about the *sum of natural numbers* and its diagrammatic proof are taken from Nelsen 1993, page 69. Figure 47 shows the theorem and its diagrammatic proof.

$$\frac{n \times (n+1)}{2} = 1+2+3+\cdots+n$$

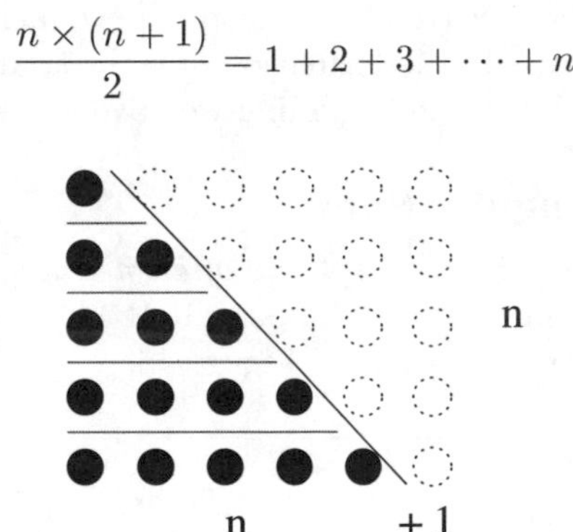

FIGURE 47 Sum of all natural numbers.

A diagrammatic proof starts by taking an n by $n+1$ rectangle. Cut it down the diagonal so that two identical isosceles triangles whose sides are of length n are formed. Now, take one of the triangles and split a side from it. Continue splitting sides until a triangle is exhausted. Note that in this way one gets the enumeration of natural numbers forming a triangle, and one triangle is half of the enumeration of points of the rectangle. Note also that we apply operations to both sides of the equality.

A.5 Euler's Theorem

Euler's theorem about polyhedra states that:

$$V - E + F = 2$$

where V is the number of vertices, E is the number of edges, and F is the number of faces of a polyhedron. The example is taken from Lakatos 1976 and Gamow 1988, pages 47-48. The diagrammatic proof of this theorem, which is historically due to Cauchy, is demonstrated in Figure 48 and is taken from from Lakatos 1976, pages 7-8.

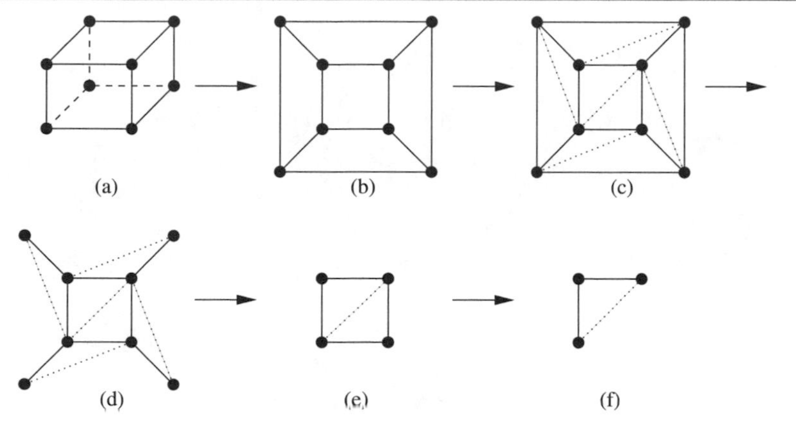

FIGURE 48 Euler's theorem.

Take any simple polyhedron (note that in our case, we take a cube, but the result is the same for any simple polyhedron). Imagine that it is hollow, and that its faces are made out of rubber (see (a) of the diagram above). Now, remove one face from the polyhedron, and stretch the rest of the polyhedron onto the plane (see (b) of the diagram). Note that since we have taken one face off, our formula should be $V - E + F = 1$. Note also that the relations between the vertices, edges and faces are preserved in this way. Triangulate all of the faces of this plane network (i.e., we are adding the same number of edges and faces to the network, so the formula remains the same — see (c) of the diagram). Now, start removing the boundary edges (see (d) of the diagram). This will have the effect of removing an edge and a face from the network at the same time, or two edges, one vertex and one face, so our formula is still preserved. We continue removing edges in appropriate order (see (e)), thus preserving the formula, until we are left with one triangle only. Clearly, for this triangle $V - E + F = 1$ holds, since there are three vertices, three edges and one face.

Notice that this diagrammatic proof is erroneous when applied to any polyhedron. The reader is referred to Lakatos 1976 for a number of counterexamples to this theorem and its proof. We describe it because it helps us to analyze various kinds of diagrammatic proofs in order to define a problem domain in §3.5. The erroneous diagrammatic schematic proof is also of interest in the discussion about the psychological validity of schematic proofs addressed in §4.5. The theorem holds for all *simple* polyhedra (Lakatos 1976, page 34).

Appendix B: The ω-Rule

Let us define the ω-rule for natural numbers as in Sundholm 1983 (note that s is the successor function – its definition can be found in Glossary):

Definition 5 (ω-Rule)
The ω-rule allows inference of the sentence $\forall x.P(x)$ from an infinite sequence $P(n)$ for $n \in \omega$ (where ω is an ordered set of natural numbers) of sentences:

$$\frac{P(0), P(s(0)), P(s(s(0))), \ldots}{\forall n.P(n)}$$

The ω-rule in an alternative to mathematical induction, in fact, it is a stronger logical rule, which is explained below. Within automated deduction, there are several motivations for using the ω-rule instead of the rule of mathematical induction – we discuss them next. Then we show the problem of using the ω-rule within implementations, and propose a solution to this problem.

B.1 Logicians' Motivation for Using ω-rule

The diagrams which represent our theorems of Category 2 are a form of representation for natural numbers. Hence, we need to formalize a theory of diagrams which is equivalent to at least a part of natural number arithmetic and is suitable for automation. Important and desirable properties of such a theory and the formalized logic are consistency, soundness and completeness. Only systems that axiomatize mathematics strongly enough may have such properties. There are two main reasons for using the ω-rule in the formalization of a theory of diagrams. The first one is that the Peano axioms plus the ω-rule form a complete theory (Orey 1956), and the second reason is that the use of the ω-rule eliminates the need for the cut rule (Prawitz 1971). The cut rule used in Gentzen's formalization of sequent calculus is as follows:

$$\frac{A, \Gamma \vdash C \quad \Gamma \vdash A}{\Gamma \vdash C}$$

The cut rule enables one to prove C using A. A is referred to as the *cut formula.* A is then eliminated by proving it from Γ.

Gödel's first incompleteness theorem says that for any formal theory of natural number arithmetic there will always be true statements for it, that are not theorems of this theory (Gödel 1931). Hence we can never completely formalize all truths of arithmetic. The usual formalization of arithmetic using Peano axioms and an induction rule is limited since Gödel's first incompleteness theorem applies to this formalization. However, Shoenfield (1959) showed that a *complete* formalization of arithmetic *can* be constructed from Peano axioms and the ω-rule, thus Gödel's incompleteness theorem does not apply here. Peano axioms plus the ω-rule is a semi-formal system because the proofs are infinite, and is therefore not a formal system in Gödel's sense (Orey 1956).

The second reason for using the ω-rule is that it removes the need to use the cut rule. For reasons such as consistency and restriction of search space, it is a desirable property of a system that cut elimination is valid (Schwichtenberg 1977). The cut elimination theorem for predicate calculus states that every proof may be replaced by one that does not require the use of a cut rule. The theorem was proved for first order logic by Gentzen (see reprint in Gentzen 1969) and for Peano axioms plus the ω-rule by Prawitz (1971). This has a significant impact on the search space in the automation of a reasoning system. If a proof is not cut-free, then any cut formula can be introduced to the proof, hence there is a potentially infinite branching of a search space. However, if the cut elimination theorem holds for a logical system, then any cut formula need not be used in the proof, hence branching of a search space is finite.[61]

Cut elimination is not valid for the inductive formalization of arithmetic, e.g., Peano axioms plus mathematical induction, as shown by Kreisel (1965). The problem which arises is that induction in Peano arithmetic is blocked for some theorems (e.g., the restricted version of *associativity of addition* with one universally quantified variable x, stated as $(x+x)+x = x+(x+x))$, because $P(s(x))$ cannot be given in terms of $P(x)$. Using the rules in the recursive definition of addition, $0+x = x$ and $s(x)+x = s(x+x)$, and cancellation of successor function, the follow-

[61]Notice, that even though the system of Peano axioms plus the ω-rule is cut-free, hence, it potentially has a finite search space, implementing the use of the ω-rule in a computer system still causes infinite branching in the proof due to the fact that an infinite number of premises need to be proved, as explained below.

ing equations represent the derivations from $P(s(x))$ to $P(x)$ (reasoning backwards). More precisely, all the possible pairwise combinations of the left hand side and the right hand side of the equations represent all the possible derivations from $P(s(x))$. Note that the term structure is different in the two sides of equations for the second arguments of both additions:

$$\begin{aligned}
\mathbf{P(s(x))} \equiv (s(x)+s(x))+s(x) &= s(x)+(s(x)+s(x)) \\
s(x+s(x))+s(x) &= s(x+(s(x)+s(x))) \\
s((x+s(x))+s(x)) &= s(x+s(x+s(x))) \\
(x+s(x))+s(x) &= x+s(x+s(x)) \qquad \not\in \mathbf{P(x)}
\end{aligned}$$

From a heuristic point of view, a generalized form of the theorem is required. This extends the problem to finding what this generalized formula might be. Arbitrarily finding it is an *ad hoc* approach, and potentially requires infinite branching in the search space. In the example about the *associativity of addition* just given, one possible generalization of the formula is $(x+y)+y = x+(y+y)$. For reasons such as these, automatic theorem proving using the usual formalization of arithmetic, i.e., Peano axioms plus mathematical induction, is very difficult. A solution may be to embed the arithmetic in a stronger system, where there is no need for generalization. An example of such a system is Peano arithmetic plus the ω-rule.

One way of putting the ω-rule into effect is to require that there is a formalization of the derivation which proves each premise. For example, one could code proofs on numbers by means of a recursive function which generates them. Such a formalization would be constructive. However, the rule as it is stated above is not constructive, and it is not suitable for implementation, since it has an infinite number of premises. Proofs with an infinite number of premises are clearly difficult to automate on a computer.

B.2 Example of Using the ω-rule

Take, for example, a special version of a theorem regarding the *associativity of addition*:

$$\forall x \ \ (x+x)+x = x+(x+x)$$

As seen in §B.1 the inductive proof is blocked, so some sort of generalization is required. In such a case the correct proof is difficult to find automatically. However, the proof can be found using the ω-rule, given that the proofs of the following premises can be generated:

$$\begin{aligned}(0+0)+0 &= 0+(0+0)\\ (s(0)+s(0))+s(0) &= s(0)+(s(0)+s(0))\\ (s(s(0))+s(s(0)))+s(s(0)) &= s(s(0))+(s(s(0))+s(s(0)))\\ &\vdots\end{aligned}$$

We restrict the ω-rule so that the infinitary proofs which are needed possess some important properties of finite proofs. One such restriction is the so called constructive ω-rule, which is explained in detail in §4.2.

Surprisingly perhaps, the formalization of arithmetic using Peano axioms and the constructive ω-rule in place of mathematical induction has the property of cut elimination and is known to be complete (Shoenfield 1959).

Glossary

Abstract Diagram

An abstract diagram is a general diagram given for some general value of magnitude, and uses abstraction devices (cf. abstraction device) such as ellipsis to represent the generality of the diagram (cf. concrete diagram).

Abstraction

Abstraction or abstraction mechanism is sometimes referred to as inductive inference, inductive learning (cf. learning induction) or generalization. It is a process of constructing a general argument from its examples.

In this book it refers to constructing a schematic proof from example-proofs. Another meaning of abstraction in this book is to refer to an *abstraction device*, such as ellipsis (cf. abstraction device), to represent general diagrams (cf. abstract diagram).

Abstraction Device

An abstraction device is a tool to represent a continuation of some pattern and is often used in an object to represent its generality. Examples of an abstraction devices include ellipsis, i.e., the "…" notation, and the summation sign *sum*.

Concrete Diagram

A concrete diagram is an instance of an abstract diagram, and is given for some particular values. No abstraction devices are needed to represent it (cf. abstract diagram).

Dependency Function

A dependency function is a linear function which by instantiation generates a natural number. This natural number indicates how many times a geometric operation is applied to the same instance of a diagram.

Diagrammatic Inference Steps

Diagrammatic inference steps are the geometric operations applied to a diagram. Chains of diagrammatic inference rules, specified by the schematic proof, form the diagrammatic proof of a theorem.

Diagrammatic Proof

A diagrammatic proof is a schematic proof that consists of applications of diagrammatic operations, i.e., diagrammatic inference rules, and this schematic proof has been checked to be correct (cf. schematic proof).

Example-Proof

An example-proof is a proof of an instance of a theorem of natural number arithmetic. It is an instance of a general diagrammatic proof. It is a list of operations applied in the proof. Several example-proofs are used to construct a general diagrammatic proof.

General Diagram

See abstract diagram.

Generalization

Generalization replaces a formula by a more general one. For example, constants, functions or predicates can be replaced by variables (e.g., $x + 3 = y$ is generalized to $x + a = y$ where 3 is a constant, and a, x and y are variables), or universally quantified variables are decoupled (e.g., $\forall x.(x + x) + x = x + (x + x)$ is generalized to $\forall x \forall y \forall z.(x + y) + z = x + (y + z)$).

x-Homogeneous Proofs

An x-homogeneous proof is a schematic proof for which there are x cases of the proof, for all values less or equal to x. The proof is x-homogeneous if all instances of the proof (for instances of numbers that are equal modulo x) have the same structure and can be abstracted to a schematic proof. All x cases need to be defined to have the smallest complete definition of a general diagrammatic proof. For example, a 2-homogeneous proof has two cases, one for the even numbers and one for the odd numbers.

Instance of Proof

See example-proof.

Instance of Theorem

An instance of a theorem is an instantiation of a universally quantified theorem for a particular value of a quantifier. For example, an instance of the theorem $n^3 = \sum_{i=0}^{n} Hex(i)$ for which $n = 4$ is $4^3 = \sum_{i=0}^{4} Hex(i) = Hex(0)+Hex(1)+Hex(2)+Hex(3)+Hex(4)$.

Instantiation

Instantiation is a process of replacing a variable with some value. Instantiation of a function is a process of assigning values to the arguments of the function and evaluating the function for these values. For example, instantiating the function Hex (for recursive definition of Hex, see recursive function) for 3 gives $Hex(3) = Hex(2+1) = Hex(2) + (6 \times 2) = (Hex(1) + (6 \times 1)) + 12 = 1 + 6 + 12 = 19$.

Internal Representation

An internal representation of a diagram is a structure, or a data type used internally on the computer to represent a diagram.

Learning Induction

Learning induction is a process of learning, i.e., inducing a new fact from examples.

Mathematical induction

Mathematical induction or standard induction is a rule of inference in some logical theory which makes an assertion about an object-level statements (cf. meta-induction). For example, in Peano arithmetic, the rule of induction is:

$$\frac{P(0) \quad P(n) \rightarrow P(s(n))}{\forall n. P(n)}$$

Meta-Induction

Meta-induction is a rule of inference in some logical theory which makes an assertion about proofs rather than object-level statements (cf. mathematical induction). For example, in Peano arithmetic, the rule of meta-induction is (where *proof* of a recursive function, and ":" stands for "is a proof of"):

$$\frac{proof(0) : P(0) \quad proof(n) : P(n) \rightarrow proof(s(n)) : P(s(n))}{\forall n. proof(n) : P(n)}$$

Meta-Level Statement

A meta-level statement is a statement about an object-level statement, in some logical theory (cf. object-level statement).

Multiple Representation

A multiple representation of a diagram is a collection of different ways of viewing the same diagram. For instance, a square can be viewed as a collection of columns or as a collection of rows.

Object-Level Statement

An object-level statement is a well-formed term, proof or inference step of the logic in use (cf. meta-level statement).

Operations

See diagrammatic inference steps.

Proof Method

A proof method is a partial specification of a tactic. Applying a method to a goal generates a list of subgoals that need to be proved. A method specifies the proof steps that the tactic performs to construct an object-level proof.

Proof Plan

Proof plan is an abstract proof specification consisting of methods which need to be applied to get an object-level proof. A proof plan is found by proof planning.

Proof Planning

Proof planning is a technique for finding proofs for mathematical theorems. The possible operators available at any stage are restricted to a set of tactics, whose preconditions are specified as methods (for more information see Bundy 1988).

Proof Tactic

A proof tactic is a program whose execution carries out part of a proof. It consists of a sequence of inference rules in some proof checking system.

Recursive Function

Recursive function is a function whose definition appeals to itself without an infinite regression. For example, Hex is a recursive function which for each input natural number x gives the xth hexagonal number:

$$\begin{aligned} Hex(0) &= 0 \\ Hex(1) &= 1 \\ Hex(n+1) &= Hex(n) + 6 \times n \end{aligned}$$

Rippling

Rippling is a process of rewriting formulae using special annotations (for more information see Bundy et al. 1993).

Schematic Proof

A schematic proof is a recursive function describing a proof of some proposition $P(n)$ in terms of n. A diagrammatic schematic proof specifies the geometric operations which need to be applied in the proof.

Standard Induction

See mathematical induction.

Successor Function

Successor function is a function that adds one to its argument. For example, $s(s(0)) = s(1) = 2$.

Symbolic (Logical) Proof

A symbolic proof is a proof in some logical theory consisting of chains of logical formulae of this theory. The proofs starts from some axioms and applies the chain of formulae to the axioms to arrive at the statement of the theorem.

Symbolic Inference Steps

Symbolic inference steps are logical rewrite formulae (cf. diagrammatic inference steps) used in a symbolic (logical) proof.

References

Amarel, S. 1968. On representations of problems of reasoning about actions. In D. Michie, ed., *Machine Intelligence 3*, pages 131–171. Edinburgh, UK: Edinburgh University Press.

Anderson, J.R. and P.J. Kline. 1979. A learning system and its psychological implications. In B. Buchanan, ed., *Proceedings of the 6th IJCAI*, pages 16–21. International Joint Conference on Artificial Intelligence, San Francisco, CA: Morgan Kaufmann.

Anderson, M., ed. 1997. *AAAI Fall Symposium on Reasoning with Diagrammatic Representations II: Working Notes*, Cambridge, MA. American Association for Artificial Intelligence, AAAI Press.

Anderson, M., P. Cheng, and V. Haarslev, eds. 2000. *Theory and Application of Diagrams: First International Conference, Diagrams 2000, Proceedings*, no. 1889 in Lecture Notes in Artificial Intelligence, Berlin, Germany. Springer Verlag.

Anderson, M. and R. McCartney. 1995. Inter-diagrammatic reasoning. In C. Mellish, ed., *Proceedings of the 14th IJCAI*, vol. 1, pages 878–884. International Joint Conference on Artificial Intelligence, San Francisco, CA: Morgan Kaufmann.

Anderson, M., B. Meyer, and P. Oliver, eds. 2001. *Diagrammatic Representation and Reasoning*. Berlin, Germany: Springer Verlag.

Bailin, S.C. and D. Barker-Plummer. 1993. Z-match: An inference rule for incrementally elaborating set instantiations. *Journal of Automated Reasoning* 11(3):391–428.

Baker, S. 1993. *Aspects of the Constructive Omega Rule within Automated Deduction*. Ph.D. thesis, University of Edinburgh, Edinburgh, UK.

Baker, S., A. Ireland, and A. Smaill. 1992. On the use of the constructive omega rule within automated deduction. In A. Voronkov, ed., *International Conference on Logic Programming and Automated Reasoning — LPAR-92*, no. 624 in Lecture Notes in Artificial Intelligence, pages 214–225. Berlin, Germany: Springer Verlag.

Baker, S. and A. Smaill. 1995. A proof environment for arithmetic with the omega rule. In J. Calmet and J. Campbell, eds., *Integrating Symbolic Mathematical Computation and Artificial Intelligence*, no. 958 in Lecture Notes in Computer Science, pages 115–130. Berlin, Germany: Springer Verlag. Also available from Edinburgh as DAI Research Paper No. 645.

Barker-Plummer, D. and S.C. Bailin. 1992. Proofs and pictures: Proving the diamond lemma with the GROVER theorem proving system. In N. Narayanan, ed., *Working Notes of the AAAI Spring Symposium on Reasoning with Diagrammatic Representations*. American Association for Artificial Intelligence, Cambridge, MA: AAAI Press.

Barwise, J. and J. Etchemendy. 1991. Visual information and valid reasoning. In W. Zimmerman and S. Cunningham, eds., *Visualization in Teaching and Learning Mathematics*, pages 9–24. Washington, DC: Mathematical Association of America.

Barwise, J. and J. Etchemendy. 1994. *Hyperproof*. Stanford, CA: CSLI Press. Distributed by Cambridge University Press.

Bauer, M.A. 1979. Programming by examples. *Artificial Intelligence* 12:1–21.

Biermann, A.W. 1972. On the inference of turing machines from sample computations. *Artificial Intelligence* 3:181–198.

Blackwell, A. 1997. Thinking with diagrams workshop: Final report. Available on web: `http://www.mrc-apu.cam.ac.uk/personal/-alan.blackwell/Workshop.html`.

Boyer, R.S. and J.S. Moore. 1979. *A Computational Logic*. London, UK: Academic Press. ACM monograph series.

Boyer, R.S. and J.S. Moore. 1990. A theorem prover for a computational logic. In M. Stickel, ed., *10th Conference on Automated Deduction*, no. 449 in Lecture Notes in Artificial Intelligence, pages 1–15. Berlin, Germany: Springer Verlag.

Brachman, R.J. and H.J. Levesque, eds. 1985. *Readings in Knowledge Representation*. San Francisco, CA: Morgan Kaufmann.

Bundy, A. 1983. *The Computer Modelling of Mathematical Reasoning*. London, UK: Academic Press. Second Edition.

Bundy, A. 1988. The use of explicit plans to guide inductive proofs. In E. Lusk and R. Overbeek, eds., *9th Conference on Automated Deduction*, no. 310 in Lecture Notes in Computer Science, pages 111–120. Berlin, Germany: Springer Verlag. Longer version available from Edinburgh as DAI Research Paper No. 349.

Bundy, A., A. Stevens, F. van Harmelen, A. Ireland, and A. Smaill. 1993. Rippling: A heuristic for guiding inductive proofs. *Artificial Intelligence* 62:185–253. Also available from Edinburgh as DAI Research Paper No. 567.

Bundy, A., F. van Harmelen, J. Hesketh, and A. Smaill. 1991. Experiments with proof plans for induction. *Journal of Automated Reasoning* 7:303–324. Earlier version available from Edinburgh as DAI Research Paper No 413.

Bundy, A., F. van Harmelen, C. Horn, and A. Smaill. 1990. The Oyster-Clam system. In M. Stickel, ed., *10th Conference on Automated Deduction*, no. 449 in Lecture Notes in Artificial Intelligence, pages 647–648. Berlin, Germany: Springer Verlag. Also available from Edinburgh as DAI Research Paper No. 507.

Burnett, M.M. and M.J. Baker. 1994. A classification system for visual programming languages. *Journal of Visual Languages and Computing* pages 287–300.

Chandrasekaran, B., J. Glasgow, and N.H. Narayanan, eds. 1995. *Diagrammatic Reasoning: Cognitive and Computational Perspectives*. Cambridge, MA: AAAI Press/MIT Press.

Chou, S.C. 1988. *Mechanical geometry theorem prover*. Dordrecht, Holland: Reidel.

Constable, R.L., S.F. Allen, H.M. Bromley, et al. 1986. *Implementing Mathematics with the Nuprl Proof Development System*. London, UK: Prentice Hall.

DeJong, G. and R. Mooney. 1986. Explanation-based learning: An alternate view. *Machine Learning* 1(2):145–176.

Dershowitz, N. and J.-P. Jouannaud. 1990. Rewriting systems. In J. van Leeuwen, ed., *Handbook of Theoretical Computer Science*, vol. B: Formal Methods and Semantics, pages 200–213. Amsterdam, Holland: Elsevier.

Descartes, R. 1954. *The geometry of Réne Descartes*. New York: Dover Publications. Translated from the French and Latin (La géometrie 1637) by Smith, D.E. and Latham, M.L.

Dubnov, IA.S. 1963. *Mistakes in geometric proofs*. Boston, MA: Heath. Translated and adapted from Russian (Oshibki v geometricheskikh dokazatelstvakh) by Henn, A.K. and Titelbaum O.A.

Dudeney, H.E. 1958. *Amusements in Mathematics*. New York: Dover Publications.

Euler, L. 1795. *Letters of Euler to a German princess, on different subjects in physics and philosophy*, vol. 2. London, UK: Printed for the translator and for H. Murray. Translated from the French (Lettres á une Princesse d'Allemagne, 1772) by H. Hunter.

Farrow, E. 1997. *Virtual Environment Simulator for Diagrammatic Reasoning*. Msc thesis, Department of Computer Science, University of Edinburgh, Edinburgh, UK.

Funt, B.V. 1980. Problem-solving with diagrammatic representations. *Artificial Intelligence* 13:201–230. Reprinted in Chandrasekaran et al. 1995, pages 33–68.

Furnas, G.W. 1990. Formal models for imaginal deduction. In *Proceedings of the Twelfth Annual Conference of the Cognitive Science Society*, pages 662–669. Mahwah, NJ: Lawrence Erlbaum Associates.

Furnas, G. 1992. Reasoning with diagrams only. In N. Narayanan, ed., *AAAI Spring Symposium on Reasoning with Diagrammatic Representations: Working Notes*, pages 118–123. American Association for Artificial Intelligence, Cambridge, MA: AAAI Press.

Gamow, G. 1988. *One two three ... infinity: Facts & Speculations of Science.* New York: Dover Publications.

Gardner, M. 1981. *Mathematical Circus*. New York: Vintage.

Gardner, M. 1986. *Knotted Doughnuts and Other Mathematical Entertainments*. New York: W.H. Freeman and Company.

Gelernter, H. 1963. Realization of a geometry theorem-proving machine. In E. Feigenbaum and J. Feldman, eds., *Computers and Thought*, pages 134–152. New York: McGraw Hill.

Gentzen, G. 1969. *The Collected Papers of Gerhard Gentzen.* Amsterdam, Holland: North-Holland. Edited by Szabo, M.E.

Gilmore, P.C. 1970. An examination of the geometry theorem-proving machine. *Artificial Intelligence* 1:171–187.

Glasgow, J.I. and D. Papadias. 1992. Computational imagery. *Cognitive Science* 3:355–394.

Gödel, K. 1931. Über formal unentscheidbare sätze der principia mathematica und verwandter systeme i. *Monatsh. Math. Phys.* 38:173–198. English translation in Heijenoort 1967.

Goldstein, I. 1973. Elementary geometry theorem proving. AI Memo 280, Massachusetts Institute of Technology, Cambridge, MA.

Hammer, E.M. 1995. *Logic and visual information*. Stanford, CA: CSLI Press. Distributed by Cambridge University Press.

Hayes, P. 1974. Some problems and non-problems in representation theory. In *Proceedings of the 1974 AISB Summer Conference*. Society for the Study of Artificial Intelligence and Simulation of Behaviour, Brighton, UK. Reprinted in Brachman and Levesque 1985, pages 4–21.

Hegarty, M. and M.A. Just. 1993. Constructing mental models of amchines from text and diagrams. *Journal of Memory and Language* 32:717–742.

Heijenoort, J van. 1967. *From Frege to Gödel: a source book in Mathematical Logic, 1879-1931*. Harvard, MA: Harvard University Press.

Iwasaki, Y., S. Tessler, and K.H. Law. 1995. Qualitative structural analysis through mixed diagrammatic and symbolic reasoning. In J. Glasgow, N. Narayanan, and B. Chandrasekaran, eds., *Diagrammatic Reasoning: Cognitive and Computational Perspectives*, chap. 21, pages 711–729. Cambridge, MA: AAAI Press/MIT Press.

Jamnik, M. 1999. *Automating Diagrammatic Proofs of Arithmetic Arguments*. Ph.D. thesis, Division of Informatics, University of Edinburgh, Edinburgh, UK.

Jamnik, M., A. Bundy, and I. Green. 1997a. Automation of diagrammatic proofs in mathematics. In B. Kokinov, ed., *Perspectives on Cognitive Science*, vol. 3, pages 168–175. Sofia, Bulgaria: New Bulgarian University. Also available from Edinburgh as DAI Research Paper No. 835.

Jamnik, M., A. Bundy, and I. Green. 1997b. Automation of diagrammatic reasoning. In M. Pollack, ed., *Proceedings of the 15th IJCAI*, vol. 1, pages 528–533. International Joint Conference on Artificial Intelligence, San Francisco, CA: Morgan Kaufmann. Also published in the "Proceedings of the 1997 AAAI Fall Symposium". Also available from Edinburgh as DAI Research Paper No. 873.

Jamnik, M., A. Bundy, and I. Green. 1998. Verification of diagrammatic proofs. In B. Meyer, ed., *Proceedings of the 1998 AAAI Fall Symposium on Formalising Reasoning with Visual and Diagrammatic Representations*, pages 23–30. American Association for Artificial Intelligence, Cambridge, MA: AAAI Press. Also published in the Proceedings of "Thinking With Diagrams 1998" Workshop. Also available from Edinburgh as DAI Research Paper No. 924.

Jamnik, M., A. Bundy, and I. Green. 1999. On automating diagrammatic proofs of arithmetic arguments. *Journal of Logic, Language and Information* 8(3):297–321.

Johnson-Laird, P.N. 1983. *Mental Models: Towards a Cognitive Science of Language, Inference and Consciousness*. Cambridge, UK: Cambridge University Press.

Kaufman, S.G. 1991. A formal theory of spatial reasoning. In J. Allen, R. Fikes, and E. Sandewall, eds., *Proceedings of the Second International Conference on Principles of Knowledge Representation and Reasoning, KR-91*, pages 347–356. San Francisco, CA: Morgan Kaufmann.

Kempe, A.B. 1879. On the geographical problem of the four colours. *American Journal of Mathematics* 2:193–200.

Koedinger, K.R. and J.R. Anderson. 1990. Abstract planning and perceptual chunks. *Cognitive Science* 14:511–550. Reprinted in Chandrasekaran et al. 1995, pages 577–625.

Kosslyn, S.M. 1993. Images in the computer and images in the brain. *Computational Intelligence* 9(4):340–342.

Kreisel, G. 1965. Mathematical logic. In T. Saaty, ed., *Lectures on Modern Mathematics*, vol. 3, pages 95–195. New York: Wiley.

Kulpa, Z. 1994. Diagrammatic representation and reasoning. *Machine Graphics and Vision* 3(1/2):77–103.

Lakatos, I. 1976. *Proofs and Refutations: The Logic of Mathematical Discovery*. Cambridge, UK: Cambridge University Press.

Larkin, J.H. and H.A. Simon. 1987. Why a diagram is (sometimes) worth ten thousand words. *Cognitive Science* 11:65–99. Reprinted in Chandrasekaran et al. 1995, pages 69–109.

Lemon, O., M. de Rijke, and A. Shimojima, eds. 1999. *Journal of Logic, Language and Information,* **8**(3). Dordrecht, Holland: Kluwer. Special Issue on Efficacy of Diagrammatic Reasoning.

Lindsay, R.K. 1998. Using diagrams to understand geometry. *Computational Intelligence* 14(2):222–256.

Lindsay, R.K. 2000. Playing with diagrams. In M. Anderson, P. Cheng, and V. Haarslev, eds., *Theory and Application of Diagrams: First International Conference, Diagrams 2000, Proceedings*, no. 1889 in Lecture Notes in Artificial Intelligence, pages 300–313. Berlin, Germany: Springer Verlag.

Lucas, J.R. 1970. *The Freedom of the Will*. Oxford: Clarendon Press.

MacLane, S. 1971. *Categories for the Working Mathematician*. Berlin, Germany: Springer Verlag.

Manna, Z. and R.J. Waldinger. 1985. *The Logical Basis for Computer Programming, Vol 1: Deductive Reasoning*. Reading, MA: Addison-Wesley.

Maxwell, E.A. 1959. *Fallacies in Mathematics*. Cambridge, UK: Cambridge University Press.

McDougal, T.F. 1993. Using case-based reasoning and situated activity to write geometry proofs. In *Proceedings of the Fifteenth Annual Conference of the Cognitive Science Society*, pages 711–716. Mahwah, NJ: Lawrence Erlbaum Associates.

McDougal, T.F. and K.J. Hammond. 1993. Representing and using procedural knowledge to build geometry proofs. In *Proceedings of the Eleventh National Conference on Artificial Intelligence*, pages 60–65. Cambridge, MA: AAAI Press/MIT Press.

McDougal, T.F. and K.J. Hammond. 1995. Using diagrammatic features to index plans for geometry theorem-proving. In J. Glasgow, N. Narayanan, and B. Chandrasekaran, eds., *Diagrammatic Reasoning: Cognitive and Computational Perspectives*, chap. 17, pages 691–709. Cambridge, MA: AAAI Press/MIT Press.

Meyer, B., ed. 1998. *Proceedings of the 1998 AAAI Fall Symposium on Formalising Reasoning with Visual and Diagrammatic Representations*, Cambridge, MA. American Association for Artificial Intelligence, AAAI Press.

Michalski, R.S. 1983. A theory and methodology of inductive learning. *Artificial Intelligence* 20:111–161.

Mitchell, T.M. 1978. *Version Spaces: An approach to concept learning*. Ph.D. thesis, Stanford University, Stanford, CA.

Mitchell, T.M. 1982. Generalization as search. *Artificial Intelligence* 18:203–226.

Mitchell, T.M., R.M. Keller, and S.T. Kedar-Cabelli. 1986. Explanation-based generalization: A unifying view. *Machine Learning* 1(1):47–80. Also available as Tech. Report ML-TR-2, SUNJ Rutgers, 1985.

Muggleton, S.H. 1991. Inductive logic programming. *New Generation Computing* 8(4):295–318.

Narayanan, N.H., ed. 1992. *AAAI Spring Symposium on Reasoning with Diagrammatic Representations: Working Notes*, Cambridge, MA. American Association for Artificial Intelligence, AAAI Press.

Nelsen, R.B. 1993. *Proofs without Words: Exercises in Visual Thinking*. Washington, DC: Mathematical Association of America.

Nelsen, R.B. 2001. *Proofs without Words II: Exercises in Visual Thinking*. Washington, DC: Mathematical Association of America.

Nevins, A. 1975. Plane geometry theorem-proving using forward chaining. *Artificial Intelligence* 6:1–23.

Newell, A. and H.A. Simon. 1956. The Logic Theory machine: A complex information processing system. *IRE Transactions on Information Theory* 2(3):61–79.

Nicholas, J.M., ed. 1977. *Images, Perception, and Knowledge*. Dordrecht, Holland: Reidel.

Olivier, P., ed. 1996. *AAAI Spring Symposium on Cognitive & Computational Models of Spatial Representation: Working Notes*, Cambridge, MA. American Association for Artificial Intelligence, AAAI Press.

Orey, S. 1956. On ω-consistency and related properties. *Journal of Symbolic Logic* 21:246–252.

Paulson, L.C. 1989. The foundation of a generic theorem prover. *Journal of Automated Reasoning* 5:363–397.

Paulson, L.C. 1991. *ML for the Working Programmer*. Cambridge, UK: Cambridge University Press.

Penrose, R. 1989. *The Emperor's New Mind*. New York: Vintage.

Penrose, R. 1994a. Mathematical intelligence. In J. Khalfa, ed., *What is Intelligence?*, pages 107–136. The Darwin College Lectures, Cambridge, UK: Cambridge University Press.

Penrose, R. 1994b. *Shadows of the Mind*. Oxford, UK: Oxford University Press.

Pinker, S. 1985. Visual cognition: an introduction. In S. Pinker, ed., *Visual Cognition*, pages 1–63. Cambridge, MA: MIT Press. Reprinted from Cognition: international journal of cognitive psychology, volume 18.

Plotkin, G. 1969. A note on inductive generalization. In D. Michie and B. Meltzer, eds., *Machine Intelligence 5*, pages 153–164. Edinburgh, UK: Edinburgh University Press.

Plotkin, G. 1971. A further note on inductive generalization. In D. Michie and B. Meltzer, eds., *Machine Intelligence 6*, pages 101–126. Edinburgh, UK: Edinburgh University Press.

Poincaré, H. 1899. Complément à l'analysis situs. *Rendiconti del Circolo Matematico di Palermo* 13:285–343.

Pólya, G. 1945. *How to solve it*. Princeton, NJ: Princeton University Press.

Pólya, G. 1965. *Mathematical Discovery*. New York: Wiley. Two volumes.

Prawitz, D. 1971. Ideas and results in proof theory. In J. Fenstad, ed., *Studies in Logic and the Foundations of Mathematics: Proceedings of the Second Scandinavian Logic Symposium*, vol. 63, pages 235–307. Amsterdam, Holland: North-Holland.

Pylyshyn, Z.W. 1973. What the mind's eye tells the mind's brain: A critique of mental imagery. *Psychological Bulletin* 80:1–24. Reprinted in Nicholas 1977.

Pylyshyn, Z.W. 1981. The imagery debate: Analogue media versus tacit knowledge. *Psychological Review* 88(1):16–45.

Quinlan, J.R. 1986. Induction of decision trees. *Machine Learning* 1(1):81–106.

Schwichtenberg, H. 1977. Proof theory: Some applications of cut-elimination. In J. Barwise, ed., *Handbook of Mathematical Logic*, pages 867–896. Amsterdam, Holland: North-Holland.

Shin, S.J. 1991. An information-theoretic analysis of valid reasoning with Venn diagrams. In J. Barwise, J. Gawron, G. Plotkin, and S. Tutiya, eds., *Situation Theory and Its Applications*, vol. 2. Stanford, CA: CSLI Press. Distributed by Cambridge University Press.

Shin, S.J. 1995. *The Logical Status of Diagrams*. Cambridge, UK: Cambridge University Press.

Shoenfield, J.R. 1959. On a restricted ω-rule. *Bulletin de l'Academie Polonaise des Sciences : Serie des sciences mathematiques, astronomiques et physiques* 7:405–407.

Simon, H.A. 1996. *The Sciences of the Artificial*. Cambridge, MA: MIT Press, 3rd edn.

Sloman, A. 1971. Interactions between philosophy and artificial intelligence: the role of intuition and non-logical reasoning in intelligence. *Artificial Intelligence* 2:209–225.

Sloman, A. 1996. Towards a general theory of representations. In D. Peterson, ed., *Forms of Representation: an interdisciplinary theme for cognitive science*, pages 118–140. Bristol, UK: Intellect.

Sowa, J. 1984. *Conceptual Structures: Information Processing in Mind and Machine*. Reading, MA: Addison-Wesley.

Stenning, K. and J. Oberlander. 1992. Implementing logics in diagrams. In N. Narayanan, ed., *AAAI Spring Symposium on Reasoning with Diagrammatic Representations: Working Notes*, pages 91–95. American Association for Artificial Intelligence, Cambridge, MA: AAAI Press.

Stenning, K. and J. Oberlander. 1995. A cognitive theory of graphical and linguistic reasoning: Logic and implementation. *Cognitive Science* 19:97–140.

Sundholm, B.G. 1983. *A Survey of the Omega Rule*. Ph.D. thesis, University of Oxford, Oxford, UK.

Van Baalen, J. 1989. *Toward a theory of representation design*. Ph.D. thesis, Massachusetts Institute of Technology, Cambridge, MA.

Welch, B.B. 1995. *Practical Programming in Tcl and Tk*. London, UK: Prentice Hall.

Winston, P.H. 1975. Learning structural descriptions from examples. In P. Winston, ed., *The psychology of computer vision*. New York: McGraw Hill.

Winterstein, D., A. Bundy, and M. Jamnik. 2000. A proposal for automatic diagrammatic reasoning in continuous domains. In M. Anderson, P. Cheng, and V. Haarslev, eds., *Theory and Application of Diagrams: First International Conference, Diagrams 2000, Proceedings*, no. 1889 in Lecture Notes in Artificial Intelligence, pages 286–299. Berlin, Germany: Springer Verlag.

Zisserman, A. 1992. Notes on geometric invariance in vision. BMVC92 Tutorial.

Index